W9-BMB-016

For Katrin, Allister, and Oskar, my incredibly awesome family
—M.L.

Copyright © 2019 by Mike Lowery

All rights reserved. Published by Scholastic Inc., *Publishers since 1920*. SCHOLASTIC and associated logos are trademarks and/or registered trademarks of Scholastic Inc. • The publisher does not have any control over and does not assume any responsibility for author or third-party websites or their content. • No part of this publication may be reproduced, stored in a retrieval system, or transmitted in any form or by any means, electronic, mechanical, photocopying, recording, or otherwise, without written permission of the publisher. For information regarding permission, write to Scholastic Inc., Attention: Permissions Department, 557 Broadway, New York, NY 10012.

ISBN 978-1-338-61097-0
10 9 8 7 6 5 4 3 2   19 20 21 22 23
Printed in the U.S.A.   40 • First printing 2019
The text type was set in Gotham.
The display type was hand lettered by Mike Lowery.
Book design by Doan Buu

EVERYTHING AWESOME ABOUT DINOSAURS

AND OTHER PREHISTORIC BEASTS!

ROAR!

WRITTEN AND ILLUSTRATED BY

MIKE LOWERY

Scholastic Inc.

TABLE of CONTENTS

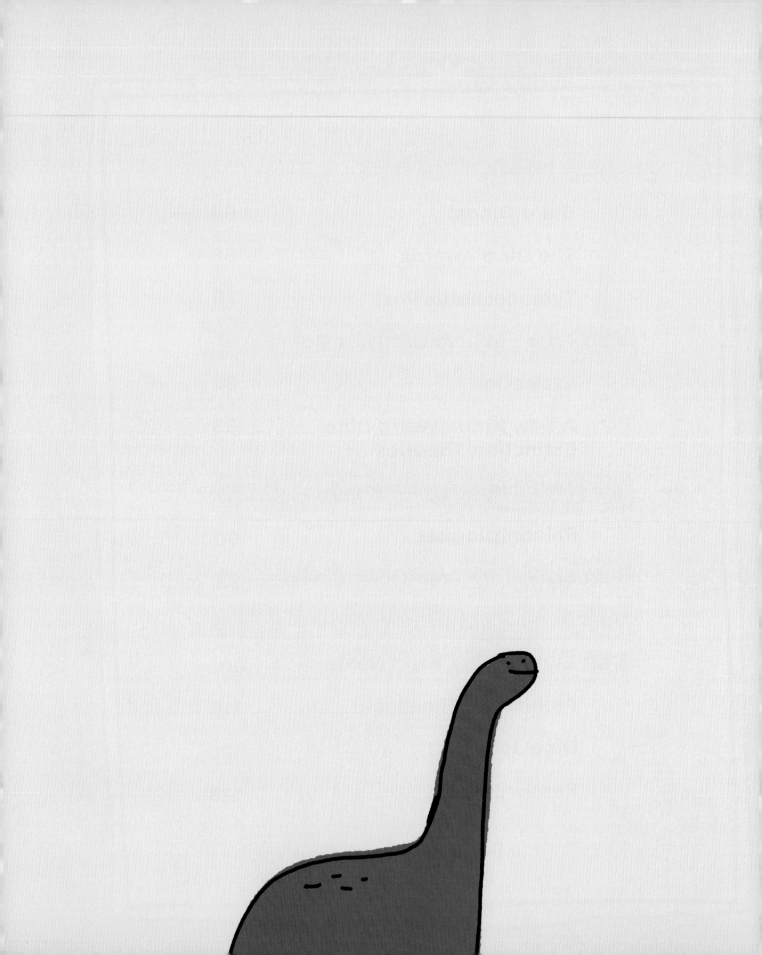

PART ONE

WHAT IS A DINOSAUR ?!

THAT'S EASY! THEY'RE... UM...

A BIG... LIZARD THING?

THE WORD DINOSAUR

COMES FROM TWO ANCIENT GREEK WORDS AND MEANS:

"TERRIBLE LIZARD"

TERRIBLE? I'M ACTUALLY REALLY NICE!

WELL, YOU DID EAT MY BROTHER...

WELL, SORT OF.

AN ENGLISH BIOLOGIST COINED THE PHRASE IN 1842, AND IT WAS IN REFERENCE TO THEIR SIZE, NOT THEIR APPEARANCE.

RICHARD OWEN

Sir Richard Owen helped start the Natural History Museum in London, tutored children of the royal family in science, and famously feuded with Charles Darwin.

Later in life, he wasn't super well liked by his peers. He has been described in one biography as being "...addicted to controversy and driven by arrogance and jealousy."

DINOSAURS WERE PRE-HISTORIC ANIMALS...

WHAT DOES "PREHISTORIC" MEAN?

GOOD QUESTION

PREHISTORIC MEANS ANYTHING THAT HAPPENED OR EXISTED IN THE TIME BEFORE WE STARTED KEEPING WRITTEN RECORDS.

The earliest forms of writing showed up roughly 5,300 years ago, but it took thousands of years for it to become widely used. Ancient Egyptians were some of the first people to keep written records.

DINO JOKES

WHAT DID THE VOLCANO SAY ON VALENTINE'S DAY?

I LAVA YOU!

WHAT IS A DINO'S LEAST FAVORITE REINDEER?

COMET

WHAT'S AS BIG AS A DINOSAUR, BUT WEIGHS NOTHING?

QUIT FOLLOWING ME!

ITS SHADOW!

NOT ALL EXTINCT ANIMALS ARE DINOSAURS.

LIKE WOOLLY MAMMOTHS.

OF COURSE I'M NOT A DINOSAUR! AND I'M NOT A LIZARD!

X NOT A DINO

Extinction is when a species dies out.

OR THE DODO.

X NOT A DINO

WHAT DID YOU CALL ME?

Lots of animals have roamed the Earth, and sadly some of them don't exist anymore. Even plant species can go extinct!

In prehistoric times, extinction was caused by: climate change, food running out, no place to live, or other animals that are better adapted.

HOW TO TELL IF IT'S A DINOSAUR:

USE THIS HANDY CHECKLIST!

✓1. DINOSAURS LIVED DURING THE MESOZOIC ERA.

THE MESOZOIC ERA LASTED FOR ABOUT 180 MILLION YEARS. THE PLANET WAS MUCH WARMER THEN.

WHEN WILL AIR CONDITIONING BE INVENTED?!

HE'S JUST KIDDING.

The Mesozoic era started roughly 251 million years ago, and lasted for about 180 million years. Dinosaurs ruled for a lot of that period, but they definitely weren't the only animals around.

☑ 2. DINOSAURS ONLY LIVED ON LAND.

THIS WASN'T A DINOSAUR!

THEN... WHAT AM I?

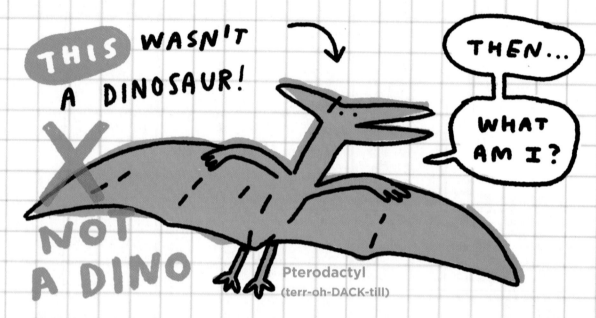

NOT A DINO

Pterodactyl
(terr-oh-DACK-till)

AND WHAT ABOUT THOSE AWESOME SWIMMING MONSTER THINGS?!

Plesiosaurus
(PLEE-see-oh-SORE-us)

NOT A DINO

☑ 3. DINOSAURS HAD BACKBONES.

Dinosaurs had spines! There's even a dinosaur named spinosaurus, which means "spine lizard." The spinosaurus had long spines on its back. These spines were called a "sail."

☑ 4. DINOSAURS WALKED WITH THEIR LEGS UNDER THEIR BODIES LIKE BIRDS.

(NOT OUT TO THE SIDES, LIKE AN ALLIGATOR).

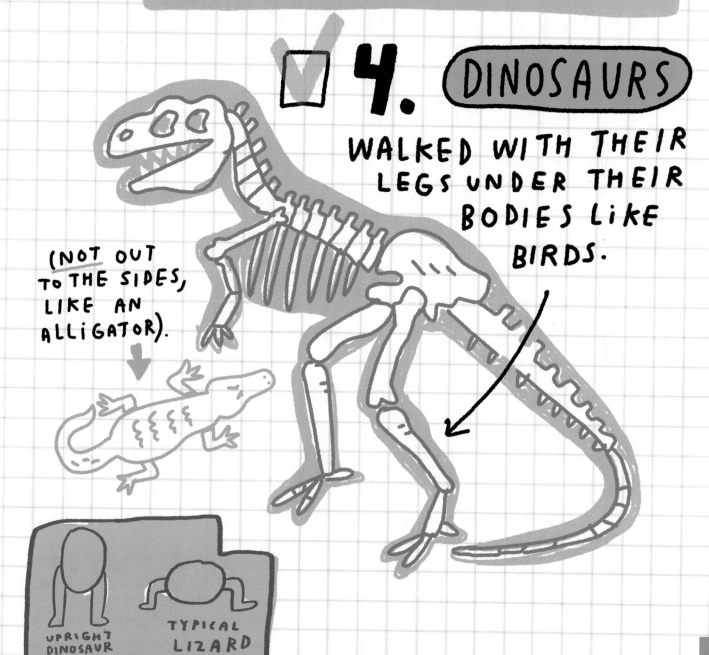

UPRIGHT DINOSAUR

TYPICAL LIZARD

SEE? I TOLD YOU THE ANSWER WAS A LITTLE COMPLICATED. ANOTHER WAY TO THINK OF IT IS LIKE THIS:

DINOSAURS*

WERE A GROUP OF PREHISTORIC, WARM-BLOODED REPTILES THAT RULED THE EARTH FOR MILLIONS OF YEARS.

SOME WERE SMALLER THAN CHICKENS.

SOME ATE MEAT.

SOME HAD HORNS!

YUMMY!

SOME WERE AS BIG AS AIRPLANES.

SOME HAD LONG NECKS.

SOME ATE VEGGIES.

AND THEY WERE ALL AWESOME!

*NOTE: MODERN-DAY BIRDS ARE ACTUALLY DINOSAURS, TOO. MORE ON THAT A LITTLE LATER IN THE BOOK.

PART TWO

A BRIEF HISTORY of EARTH!

OUR PLANET!

THE DINOSAURS DIDN'T SHOW UP UNTIL ABOUT 250 MILLION YEARS AGO AND **LOTS** OF STUFF HAPPENED BEFORE THEN.

BUT OBVIOUSLY NOTHING AS AWESOME AS ME!

FIRST OFF, YOU SHOULD KNOW THAT OUR PLANET IS REEAAAALLLLY OLD. SCIENTISTS THINK IT'S MORE THAN

4.5 BILLION YEARS OLD!

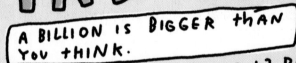

A BILLION IS BIGGER THAN YOU THINK.

- A MILLION SECONDS IS 12 DAYS.
- A BILLION SECONDS IS...

31 YEARS!

EIGHT MILLION NINE HUNDRED AND...

THAT DOES SOUND OLD.

QUICK FACT: A software developer named Jeremy Harper counted to one million and it took him around 89 days. He did it live on the internet so people could watch. He even got a Guinness World Record for highest number ever counted to by a human. If he'd kept the same pace it would take 244 YEARS to count to one billion.

IF THE HISTORY OF the EARTH WAS SMASHED INTO JUST ONE, 24-HOUR DAY, DINOSAURS WOULDN't EVEN MAKE AN APPEARANCE UNTIL ABOUT 10:56 P.M. HUMANS WOULDN'T ARRIVE UNTIL 11:58 P.M!

541 - 4,600 MILLION YEARS AGO

THIS IS THE EARLIEST AND LONGEST PERIOD OF TIME ON OUR PLANET. IN FACT, IT MAKES UP ABOUT 90% OF THE ENTIRE HISTORY OF EARTH! THE PRECAMBRIAN ERA BEGAN WHEN THE PLANET WAS FORMED 4.6 BILLION YEARS AGO, AND LASTED UNTIL 541 MILLION YEARS AGO.

EARLY ON, **ASTEROIDS BOMBARDED** THE EARTH.
(HADEAN EON)

LAND-MASSES BEGAN TO DEVELOP.
(ARCHEAN EON)

WATER FROM COMETS TURNED INTO RAIN.

MOLTEN SEAS WERE COOLED BY THE RAIN.

RODINIA
(FROM THE RUSSIAN FOR "MOTHER LAND")

THIS **SUPER-CONTINENT** WAS FORMED.
(PROTEROZOIC EON)

EARLY SIGNS OF LIFE

CYANOBACTERIA

(SIGH-an-o-BAC-teer-ee-uh)

SINGLE-CELL PHONE

(AKA BLUE-GREEN ALGAE) ARE TINY, SINGLE-CELLED BACTERIA THAT COULD TURN SUNLIGHT INTO ENERGY AND WOULD GIVE OFF OXYGEN AS A RESULT. THEY'RE 3.5 BILLION YEARS OLD, AND STILL AROUND TODAY!

DICKINSONIA

(dik-in-SO-nee-a)

POSSIBLY THE VERY FIRST ANIMAL!

(558 MILLION YEARS OLD) THESE ROUND, FLAT ANIMALS LIVED ON THE SEAFLOOR AND WERE BILATERALLY SYMMETRICAL (that MEANS BOTH HALVES WERE MIRROR IMAGES OF EACH OTHER). SOME WERE SMALL, BUT FOSSILS HAVE BEEN FOUND OF ONES THAT WERE 3 FEET LONG.

541 TO 251 MILLION YEARS AGO

DURING THIS TIME, EARTH WENT THROUGH LOTS OF CHANGES. THE GEOLOGY AND CLIMATE WERE CHANGING AND LIFE STARTED TO EVOLVE MORE QUICKLY. AT FIRST IT STARTED IN THE OCEANS, BUT SLOWLY, LIVING THINGS MADE IT ONTO DRY LAND.

FUNGUS

I NEED A TOWEL.

THE CAMBRIAN EXPLOSION

BOOM

SUDDENLY MORE COMPLEX ANIMALS APPEARED. (WELL, IT WAS "SUDDEN" IN GEOLOGICAL TERMS. IT HAPPENED OVER MILLIONS OF YEARS.)

SIZE: ABOUT ONE INCH LONG

ONE OF THE FIRST ANIMALS TO HAVE A SKULL!

THIS FISH-LIKE GUY HAD NO JAW, BUT DID HAVE 2 EYES AND A MOUTH.

HAIKOUICHTHYS

(HIGH-koo-ICK-thiss)

TRILOBITES

(TRAHY-luh-bahyt)

One of the most successful early life forms, trilobites had armor-plated bodies and were one of the first living things with the ability to see. This means they could HUNT prey like plankton.

Some could ball themselves up when they sensed danger.

There were over 20,000 species of trilobites that we know of!

WERE AROUND FOR ALMOST 300 MILLION YEARS!

QUICK FACT!

The Pahvant Ute people, from what is now Utah, would collect Cambrian trilobite fossils because they believed that they had special powers. They would wear them as protective jewelry.

DISTANT RELATIVES OF LOBSTERS AND SPIDERS.

SIZE: UP TO 28 INCHES LONG!

3 STATES HAVE PICKED THEM AS THE OFFICIAL STATE FOSSIL.

OHIO, WISCONSIN, AND PENNSYLVANIA

ORTHOCERAS

(OR-thoc-ER-as)

RELATED TO MODERN-DAY SQUIDS.

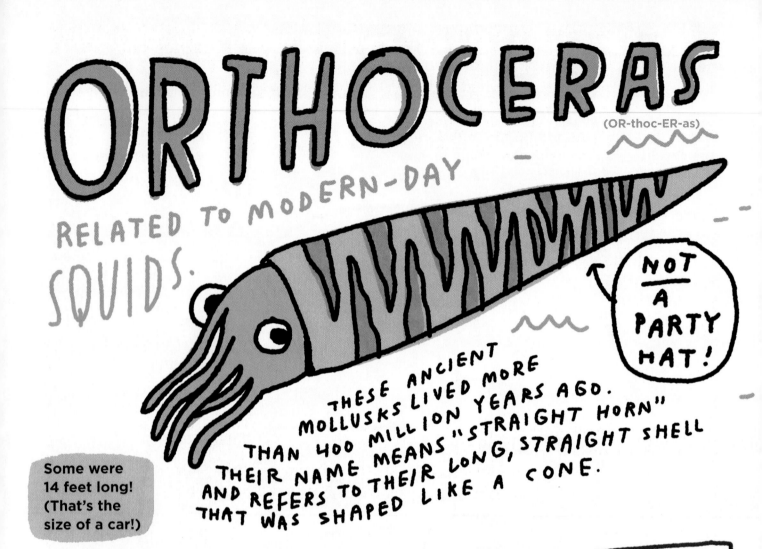

NOT A PARTY HAT!

Some were 14 feet long! (That's the size of a car!)

THESE ANCIENT MOLLUSKS LIVED MORE THAN 400 MILLION YEARS AGO. THEIR NAME MEANS "STRAIGHT HORN" AND REFERS TO THEIR LONG, STRAIGHT SHELL THAT WAS SHAPED LIKE A CONE.

PNEUMODESMUS NEWMANI

(NEW-mo-DEZ-muss NEW-mon-ee)

428 MILLION YEARS OLD

ONE OF THE FIRST AIR-BREATHING LAND ANIMALS

SIZE: LESS THAN HALF AN INCH LONG

TIKTAALIK
(tik-TAA-lik)
Tiktaalik roseae, better known as the "fishapod."
HAD A HEAD LIKE A CROCODILE
COULD GO ONTO LAND FOR SHORT PERIODS OF TIME
GILLS LIKE A FISH
Some evidence shows they may have had lungs AND gills.
SOME WERE MORE THAN 8 FEET LONG!
IS CONSIDERED TO BE A TRANSITIONAL STAGE IN EVOLUTION BETWEEN FINS AND FEET
375 MILLION YEARS OLD.

MEGANEURA: A GIANT DRAGONFLY WITH A 30-INCH WINGSPAN.

(MEG-ah-NEUR-ah)

It was a carnivore that ate other insects!

Its name means "large-veined."

– – – – – –

One of the biggest flying insects ever!

Some scientists believe that the higher oxygen at the time of *Meganeura* (compared to today) is what allowed them to grow so big!

PLANT LIFE DURING tHE PALEOZOIC ERA.

ALGAE

A very simple plant with no roots, stems, or leaves

MOSSES

Small green plant that grows like a carpet on a surface

FERNS

Seedless, nonflowering plants (usually in tropical regions)

GINKGOS

Large shade trees with fan-shaped leaves

CONIFERS

Evergreen trees and shrubs with cone-shaped seeds (like a pinecone!)

CYCADS

Plants with thick trunks and palms growing out of the top like a crown

WARNING!

THEY'RE ALMOST HERE!

THE PERIOD RIGHT BEFORE THE DINOSAURS SHOWED UP IS CALLED

THE PERMIAN PERIOD.

IT WAS the LAST PERIOD DURING the PALEOZOIC ERA. IT BEGAN ABOUT 300 MILLION YEARS AGO AND ENDED 251 MILLION YEARS AGO.

The temperature and climate during the Permian period varied a lot. It began with an ice age and cycled through periods of hot and cold.

DID SOMEBODY SAY "PERM"?!

Actually, it was named after the Russian city of Perm, where great samples of rocks from this period were found.

29

DURING THE PERMIAN PERIOD A GROUP OF MAMMAL-LIKE REPTILES CALLED THE **SYNAPSIDS** RULED THE EARTH. HERE'S ONE!

I know it kinda looks like one, but it's NOT A DINOSAUR!

DIMETRODON
(dahy-ME-truh-don)

ITS NAME MEANS "TWO MEASURES OF TEETH."

It's BELIEVED TO BE THE FIRST ANIMAL TO HAVE DIFFERENT TYPES OF TEETH.

(295-272 MILLION YEARS AGO)
THOUGH IT'S NOT A DIRECT ANCESTOR OF MAMMALS, IT'S ACTUALLY MORE CLOSELY RELATED TO MAMMALS THAN IT IS TO LIZARDS.

Scientists think some may have weighed as much as 500 pounds. That's as much as a grizzly bear!

It's believed that the "sail" on its back was either to regulate temperature OR to communicate.

WANNA GO SEE A MOVIE?

OK.

They could grow up to 15 feet. That's the distance of a free throw line on a basketball court!

ERYOPS

(EH-ree-ops)

(295 MILLION YEARS AGO)

It sorta looked like a cross between a **FROG** and an **ALLIGATOR**. RIGHT?

HAD TEETH ON THE ROOF OF ITS MOUTH!

Their heads were big! Their skulls weighed almost a third of their entire body weight.

UP TO 6.6 FEET LONG

ON LAND THEY COULD JUST LIFT THEMSELVES UP A LITTLE, WHICH MADE THEM SLOW, BUT IN THE WATER THEY WERE **FAST AND EXCELLENT HUNTERS**.

DINOCEPHALIA

(270 MILLION YEARS AGO)

(DIE-NO-sef-a-LEE-ya)

DON'T LET ITS NAME FOOL YOU! THIS DUDE ISN'T A DINOSAUR!

THEY WERE SOME OF THE BIGGEST ANIMALS IN THE PERMIAN PERIOD.

SOME HAD **THICK** SKULLS THAT MAY HAVE BEEN USED FOR HEAD BUTTING.

Weighed as much as a modern RHINO!

WEIGHED: UP TO 4,400 LBS

SIZE: UP TO 15 FEET LONG

SCUTOSAURUS

(Skoo-TOE-SORE-us)

(250 MILLION YEARS AGO)

ITS NAME MEANS "SHIELD LIZARD."

ARMOR PLATED →

HERBIVORE ↗

THEY WERE UP TO 6 FEET LONG AND 1,000 POUNDS!

THAT'S MORE THAN A PIANO.

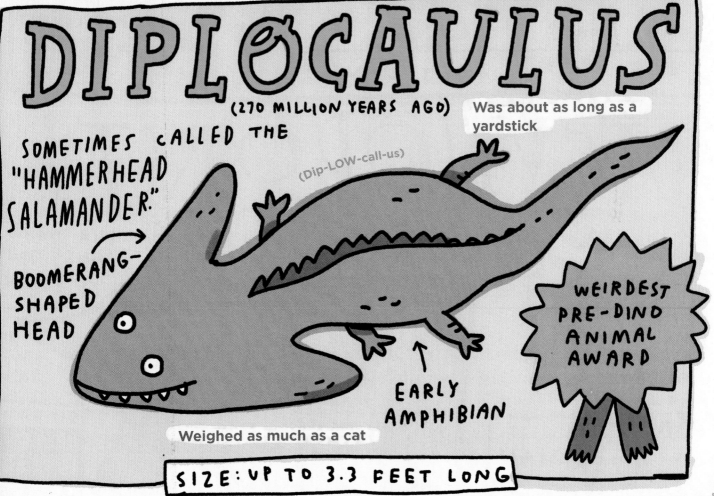

DIPLOCAULUS

(270 MILLION YEARS AGO)

(Dip-LOW-call-us)

Was about as long as a yardstick

SOMETIMES CALLED THE "HAMMERHEAD SALAMANDER."

BOOMERANG-SHAPED HEAD →

EARLY AMPHIBIAN ↑

WEIRDEST PRE-DINO ANIMAL AWARD

Weighed as much as a cat

SIZE: UP TO 3.3 FEET LONG

GORGONOPSIA

(GOR-ga-NOP-see-a)

A GROUP THAT INCLUDED SOME OF THE LARGEST **CARNIVORES** OF THE LATE PERMIAN PERIOD

About as long as a great white shark.

Weighed as much as a cow.

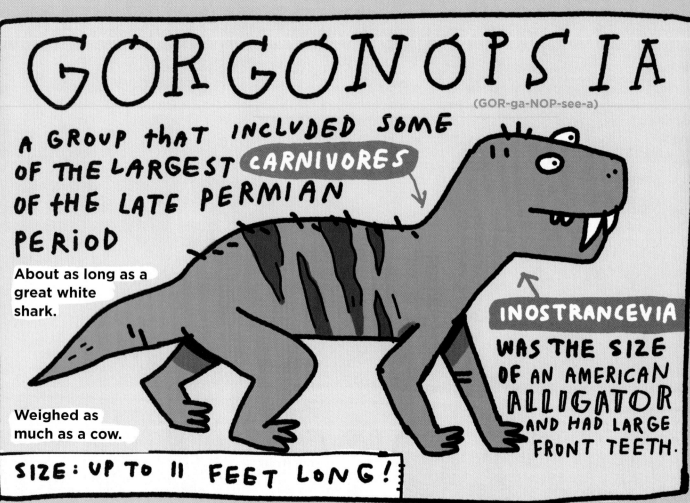

INOSTRANCEVIA WAS THE SIZE OF AN AMERICAN **ALLIGATOR** AND HAD LARGE FRONT TEETH.

SIZE: UP TO 11 FEET LONG!

DICYNODONT

(DEE-cee-NO-dant)

HERBIVORE

THEIR NAME MEANS "TWO DOG TOOTH."

SOME HAD A HARD BEAK AND TWO TUSKS.

SHORT TAIL

MID PERMIAN

DURING THE LATE PERMIAN, THEY WERE THE RULING LAND VERTEBRATES.

SIZE: UP TO 13 FEET LONG

Some survived until the late Triassic and co-existed with dinosaurs.

About the size of a modern-day sheep.

ALSO DURING THE PERMIAN PERIOD, THE SUPERCONTINENT **PANGAEA** FORMED.

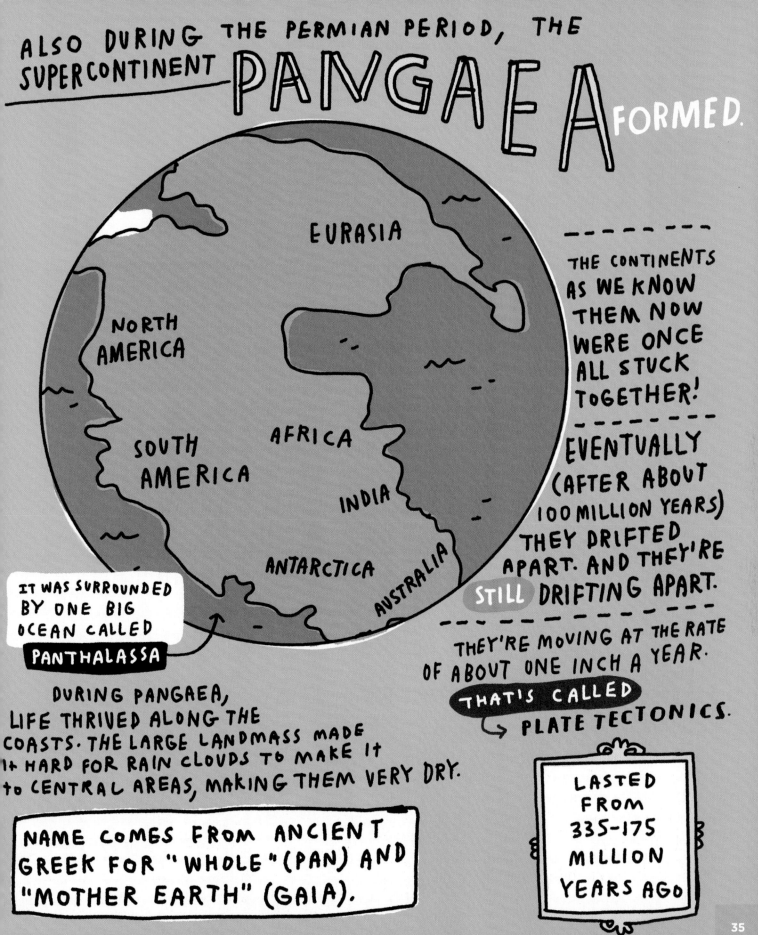

EURASIA

NORTH AMERICA

SOUTH AMERICA

AFRICA

INDIA

ANTARCTICA

AUSTRALIA

THE CONTINENTS AS WE KNOW THEM NOW WERE ONCE ALL STUCK TOGETHER!

EVENTUALLY (AFTER ABOUT 100 MILLION YEARS) THEY DRIFTED APART. AND THEY'RE STILL DRIFTING APART.

THEY'RE MOVING AT THE RATE OF ABOUT ONE INCH A YEAR. THAT'S CALLED → PLATE TECTONICS.

IT WAS SURROUNDED BY ONE BIG OCEAN CALLED PANTHALASSA

DURING PANGAEA, LIFE THRIVED ALONG THE COASTS. THE LARGE LANDMASS MADE It HARD FOR RAIN CLOUDS TO MAKE It to CENTRAL AREAS, MAKING THEM VERY DRY.

NAME COMES FROM ANCIENT GREEK FOR "WHOLE" (PAN) AND "MOTHER EARTH" (GAIA).

LASTED FROM 335-175 MILLION YEARS AGO

TOTALLY WEIRD PALEOZIC MARINE LIFE

MEGALOGRAPTUS (MAY-ga-lowg-RAP-tss) (485 MYA)

31-INCH-LONG SWIMMING SCORPIONS!

GEMUENDINA (JEM-yoo-en-DEE-na) (419 MYA)

11 INCHES LONG

HAD ARMOR PLATES AND WINGS!

TULLIMONSTRUM (ta-li-MON-strum)

AKA TULLY MONSTER (300 MYA)

THESE SMALL WEIRDOS (ABOUT 3 INCHES LONG) HAD ONE LONG CLAW-THING FOR CATCHING STUFF TO EAT.

It's the state fossil of Illinois, where they've found thousands of them!

HALLUCIGENIA (ha-lucy-JEAN-ee-a)

ONE-INCH-LONG TUBULAR ORGANISM

TOTALLY TUBULAR, DUDE!

535 MYA

Had a ring of teeth in its mouth and another set running down to its stomach!

THE GREAT PERMIAN EXTINCTION!

AKA "THE GREAT DYING"

APPROXIMATELY 252 MILLION YEARS AGO, AT THE END OF THE PERMIAN PERIOD, RIGHT BEFORE THE TRIASSIC PERIOD (WHICH IS WHEN the DINOSAURS FINALLY SHOWED UP), SOMETHING HAPPENED THAT WIPED OUT LIFE ALL OVER THE PLANET.

NEARLY	LESS THAN	UP TO
95%	5%	70%
OF ALL PLANT + ANIMAL LIFE DIED.	OF ANIMALS IN THE OCEAN SURVIVED.	OF THE LAND ANIMALS DIED.

IT'S THE WORST KNOWN EXTINCTION EVENT EVER ON EARTH. EVEN WORSE THAN THE ONE THAT KILLED THE DINOSAURS.

THAT'S SO SAD!

SNIFF

AW

EVEN MOST OF THE PLANET'S BUGS DIED. IN FACT, IT'S THE ONLY KNOWN MASS EXTINCTION EVENT FOR INSECTS.

SCIENTISTS AREN'T TOTALLY SURE WHY IT HAPPENED. IT WASN'T AN OVERNIGHT EVENT, IT GRADUALLY TOOK PLACE OVER 15 MILLION YEARS. THAT'S ACTUALLY REALLY FAST IN GEOLOGICAL TIME.

SOME SCIENTISTS BELIEVE IT WAS THE RESULT OF MASSIVE VOLCANIC ERUPTIONS THAT MADE THE AIR TOXIC AND THE OCEANS ACIDIC.

SOME SCIENTISTS BELIEVE IT WAS THE RESULT OF AN ASTEROID 4-7.5 MILES LONG HITTING EARTH. THAT'S THE SIZE OF

MOUNT EVEREST!

But life continued! It took up to 10 million years to recover, but slowly new, complex life forms began to emerge.

(251-66 MILLION YEARS AGO)

Mesozoic means "middle life" since it's between the most ancient time and today.

AKA THE AGE OF REPTILES!

IT'S ALSO KNOWN AS "THE AGE OF CONIFERS" (A TYPE OF TREE), BUT THAT DOESN'T SOUND AS COOL.

NOT AS COOL AS A DINO →

AW, MAN!

THE MESOZOIC ERA LASTED FOR ALMOST 185 MILLION YEARS AND IS DIVIDED INTO **THREE** PERIODS: **THE TRIASSIC, JURASSIC, AND CRETACEOUS.**

SOUND FAMILIAR? THAT'S ALSO WHEN **THE DINOSAURS** RULED the EARTH!

DURING THE MESOZOIC ERA, THE SUPERCONTINENT PANGAEA STARTED TO BREAK UP AND DRIFT APART.

55 MILLION YEARS AGO, INDIA AND ASIA CRASHED INTO EACH OTHER AND THE GROUND CRUMPLED UP! THAT'S HOW WE GOT THE HIMALAYAN MOUNTAINS!

SUPER COOL BONUS FACT

REMEMBER THAT THIS IS JUST AFTER THAT MASS EXTINCTION EVENT, THE GREAT PERMIAN EXTINCTION. SO, LIFE ON EARTH HAD A LOT OF RECOVERING TO DO.

BANDAGES

DINOSAURS MOST LIKELY SHOWED UP AROUND THE MIDDLE OR LATE TRIASSIC, BUT WEREN'T THE DOMINANT ANIMALS FOR A WHILE.

HERE ARE TWO OF THE FIRST KNOWN DINOSAURS.

Widely considered to be the very first dinosaur

MEAT EATER

ABOUT 40 INCHES LONG

(EE-oh-rap-tor)

EORAPTOR

Name means "Dawn Thief"

Was about the size of a beagle, and could run as fast as a rabbit

NAMED AFTER A GOAT FARMER WHO FOUND THE FIRST SPECIMEN ↓

HERRERASAURUS
(her-AYR-ah-SORE-us)
↓

MEAT EATER ↑

UP TO 20 FEET LONG!

South America might be where dinosaurs first appeared. Early dinosaurs like the herrerasaurus, eoraptor, and panphagia showed up there roughly 231 million years ago.

HERE ARE SOME NON-DINO TRIASSIC ANIMALS

THERAPSIDS

MAMMAL-LIKE REPTILES THAT SURVIVED THE PERMIAN EXINCTION

MEAT EATER)

CYNOGNATHUS

(si-no-NAY-thuss)

ABOUT 3 FEET LONG

THEY WERE ALMOST ALL OVER THE WORLD!

THERAPSIDS THRIVED DURING THE EARLY TRIASSIC, BUT BY THE MID-TRIASSIC, THEY WERE MOSTLY EXTINCT!

THRINAXODON

(thri-NAX-a-DON)

HAIRY

20 INCHES LONG

COULD BURROW INTO THE GROUND

HYPERODAPEDON
(hi-per-O-dap-PEE-don)

BEAK → (4 FEET LONG)

TANYSTROPHEUS
(tan-nis-TRO-fee-uss)

THIS REPTILE HAD A NECK THAT WAS LONGER THAN ITS BODY AND TAIL COMBINED.

SUPER-LONG NECK

It probably spent most of its time beside the riverbanks or the shoreline using its long neck to catch fish.

They were 20 feet long and 10 feet of that was neck!

Was as long as a green anaconda and weighed as much as a panda.

CHASMATOSAURUS
(kaz-ma-ta-SORE-us)

(6.5 FEET LONG)

HEAVY TAIL

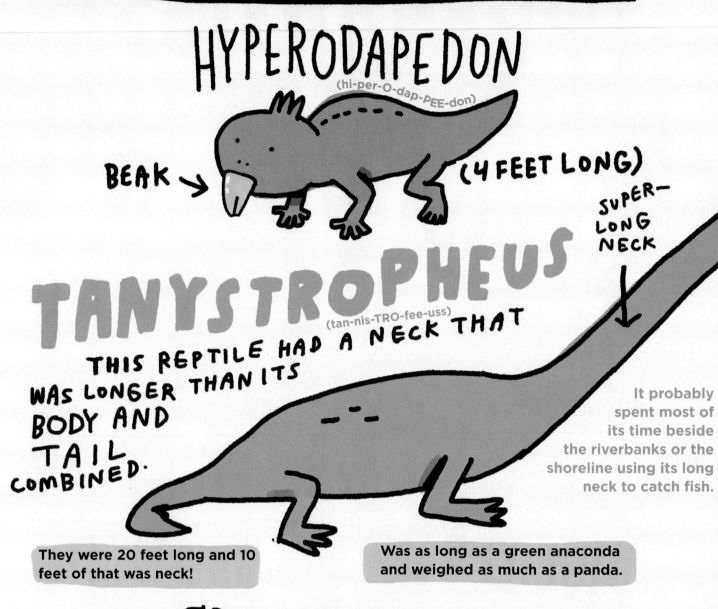

EARTH'S STRATA

By looking at the Earth's strata (the layers of rock in the ground), we can see that from the Permian to the early Triassic the planet was hot and dry. However, there was a period where it rained for a really long time. It's known as the

CARNIAN PLUVIAL EVENT.

It actually might've helped the dinosaurs diversify by causing changes in the climate and plant life.

TRIASSIC BUGS!

Many of the Earth's bugs had been wiped out during the Permian Mass Extinction, but during the Triassic period, several groups of insects appeared that we still have around today, including:

DIPTERA
(DIP-ter-uh)

BZZZ

(FLIES)

LEPIDOPTERA
(le-pi-DOP-ter-uh)

(MOTHS AND BUTTERFLIES)

HYMENOPTERA
(HI-men-nop-TER-uh)

HI

(ANCIENT ANCESTORS OF BEES, WASPS, AND ANTS)

SOME NEW TYPES OF SPIDERS ALSO APPEARED.

OPISTHOTHELAE
(op-i-SO-thee-lee)

SCORPIONS, CENTIPEDES, AND MILLIPEDES HAD ALREADY BEEN AROUND FOR MILLIONS OF YEARS, AND CONTINUED TO FLOURISH.

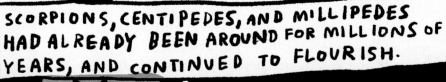

At the end of the Triassic period was ANOTHER mass extinction event, called the end-Triassic extinction (clever name, huh?). It resulted in the loss of about 76% of all life on Earth. BUT, it's also believed to be a key event that helped dinosaurs become the dominant animals on land.

NOT ANOTHER MASS EXTINCTION!

(208-145 MILLION YEAR AGO)

THE BIG DINO EXPLOSION!

This period sounds a little familiar, right? Dinosaurs may have first showed up during the Triassic, but it was during the Jurassic that they really...

BOOMED!

They continued to evolve and new dinosaurs appeared. LOTS of new dinosaurs! The Jurassic period started very warm and dry, but as the supercontinent Pangaea started to drift apart, new oceans were formed and the temperature became milder and humid.

PLATE TECTONICS

GOODBYE!

I'LL MISS YOU!

ALL THIS MOISTURE IS GREAT! WHAT A TIME TO BE ALIVE!

IT WAS WARM AND WET

LIKE A RAIN FOREST, WHICH WAS GOOD FOR PLANTS. MORE PLANTS MEANS MORE DINO FOOD!

LATER IN THE **JURASSIC** PERIOD, ALONG CAME...

BAWK!

ARCHAEOPTERYX
(ar-kee-OP-ter-ix)

(NAME MEANS "FIRST WING")

THEY'RE CONSIDERED TO BE AN IN-BETWEEN STAGE IN THE EVOLUTION BETWEEN EXTINCT DINOSAURS AND MODERN-DAY BIRDS. THIS IS KNOWN AS A

TRANSITIONAL LINK.

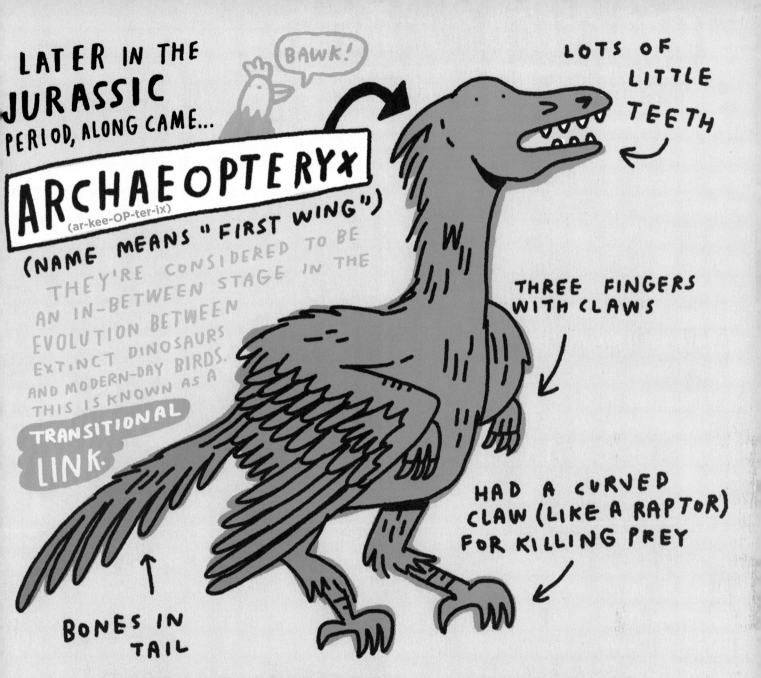

LOTS OF LITTLE TEETH

THREE FINGERS WITH CLAWS

HAD A CURVED CLAW (LIKE A RAPTOR) FOR KILLING PREY

BONES IN TAIL

They were discovered in Germany in 1860 and were called "Urvogel," which means "original bird."

JOKE TIME

WHY DID THE ARCHAEOPTERYX GET THE WORM?

BECAUSE IT WAS AN **EARLY BIRD!**

JURASSIC MARINE LIFE

The oceans during the Jurassic period were full of lots of reptiles that LOOKED like dinosaurs but weren't dinosaurs. In fact, the plesiosaurs were only distantly related to dinosaurs and were more closely related to snakes and lizards.

PLESIOSAURUS
(PLEE-see-oh-SORE-us)

LOTS OF SMALL TEETH

LARGE MARINE REPTILE

LONG NECK

Could grow up to 16 feet long!

ICHTHYOSAURUS
(ICK-thee-oh-SORE-us)

THIS DOLPHIN-LOOKING THING IS ACTUALLY A REPTILE.

AIR BREATHER

SOME HAD A DORSAL FIN.

BIG EYES

SOME LASTED INTO the LATE JURASSIC PERIOD.

THERE WERE SEVERAL MEMBERS IN the ICHTHYOSAURIA FAMILY. THE BIGGEST WAS SHINISAURUS WHO COULD GROW UP TO 15 FEET LONG.

6.5-9 FEET LONG

FIN

(GEE-oh-SORE-us)

GEOSAURUS!

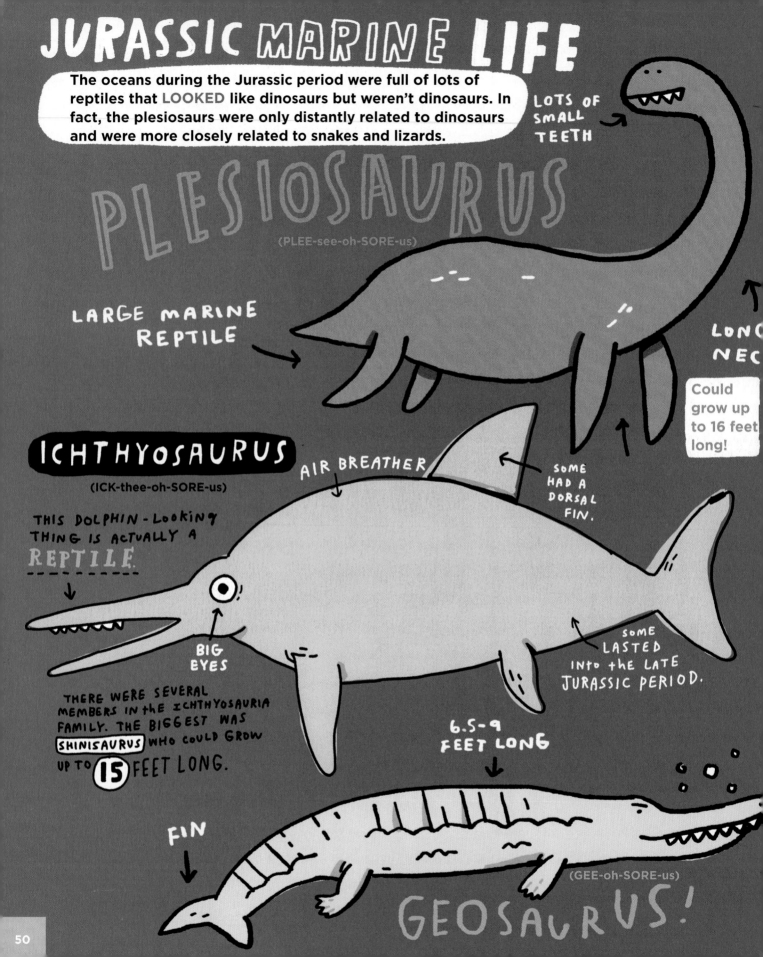

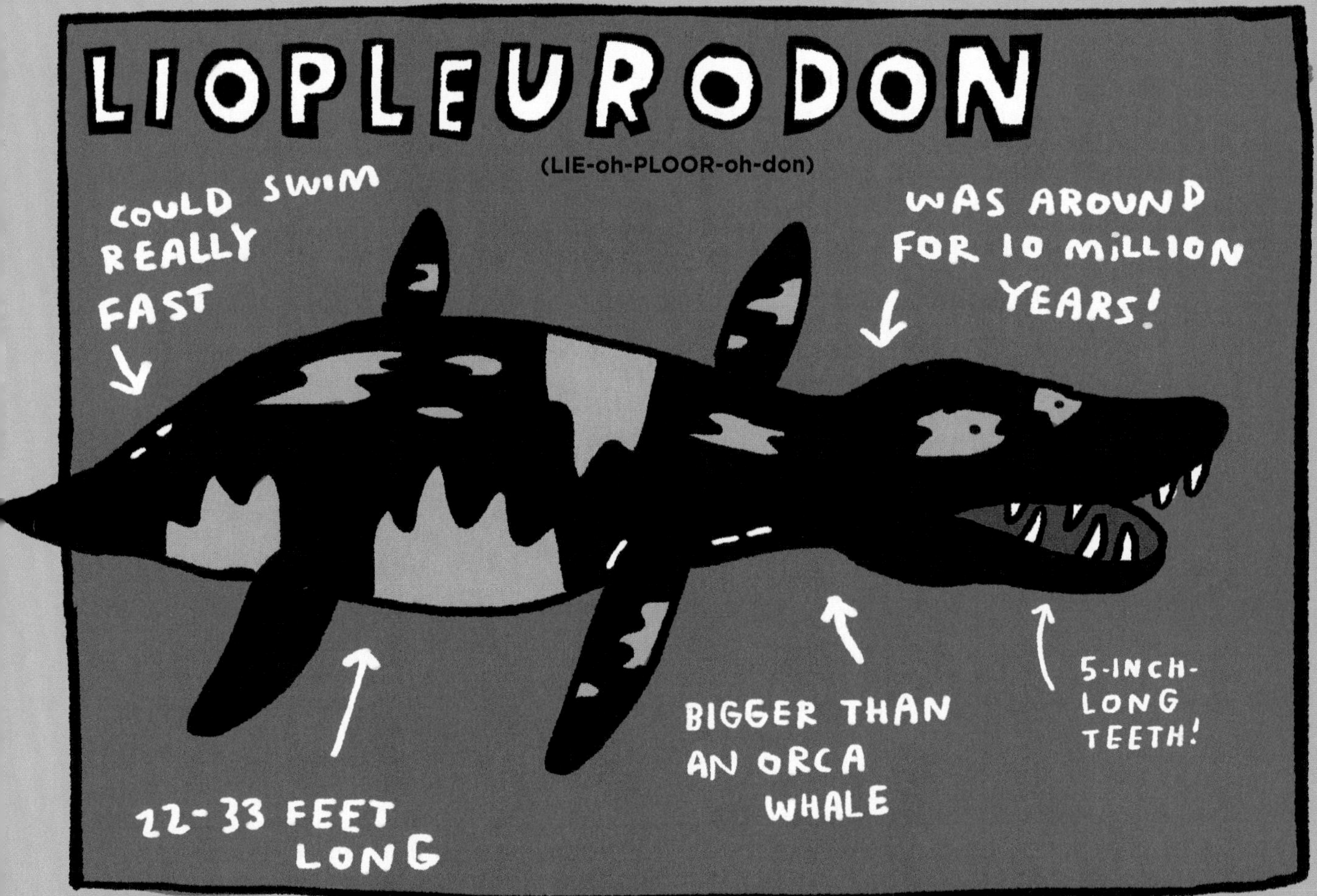

LIOPLEURODON

(LIE-oh-PLOOR-oh-don)

COULD SWIM REALLY FAST

WAS AROUND FOR 10 MILLION YEARS!

22-33 FEET LONG

BIGGER THAN AN ORCA WHALE

5-INCH-LONG TEETH!

LEEDSICHTHYS

(LEEDS-ick-thiss)

POSSIBLY THE BIGGEST FISH EVER

THEY WERE HUGE! HERE'S AN ADULT HUMAN FOR SIZE COMPARISON.

More than 50 feet long!

THE CRETACEOUS PERIOD

(145–66 MILLION YEARS AGO)

This was the LAST and LONGEST period during the Mesozoic Era.

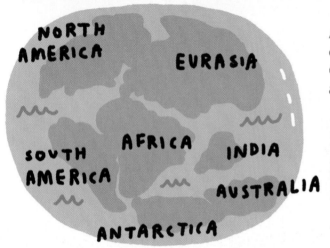

At the beginning of the Cretaceous period, the land on the planet was grouped into two big continents, Gondwana and Laurasia. By the end, it was divided and looked more like the continents we have today.

SPOILER ALERT: IT DIDN'T END SO GREAT FOR THE **DINOSAURS!**

WHAT?

Lots of awesome stuff happened during the Cretaceous period.

EARLY BIRDS STARTED TO TAKE FLIGHT.

ARCHAEOPTERYX

PLANTS STARTED TO FLOWER.

AND THIS GUY FINALLY SHOWED UP!

T. REX!

DEINOSUCHUS
(DIE-no-SOO-kuss)

This croc was 33 feet long and weighed more than TWO T. rexes! That's 15 tons!

EARLY MAMMALS

Sure, the dinosaurs were the stars of the Jurassic, but they shared the Earth with early mammals. While dinosaurs kept getting bigger during this period, mammals were mostly herbivores, insectivores, or tiny carnivores so they weren't exactly competition for the dinosaurs.

ALPHADON
(al-fa-DON)

NAME MEANS "FIRST TOOTH"

(12 INCHES LONG)

LOOKED A LITTLE LIKE AN OPOSSUM

Alphadons are related to modern-day marsupials (animals that have pouches for their young). Did you know that MOST marsupials live in Australia and New Guinea? Some of the best-known marsupials include kangaroos, wombats, koalas, possums, opossums, and Tasmanian devils!

These mammals were insectivores, which means they ate . . . insects, of course!

Speaking of EATING BUGS, there's a man in India, known in his hometown as the "insect-eating man," who entered the Guinness Book of World Records in 2003 by swallowing 200 earthworms in 30 seconds.

YUCK!

PTEROSAURS!

In case you missed it, there was a little bit of a BOMBSHELL at the beginning of this book. It turns out that...

PTERODACTYLS WEREN'T DINOSAURS.

I STILL CAN'T BELIEVE IT.

WELL, check this out, it gets even crazier. Ready?

(TELL 'EM, T. REX.)

THERE'S NO SUCH THING AS A PTERODACTYL!

UM... WHAT?!

Turns out that the word "pterodactyl" is just a really common (but wrong!) way of referring to pterodactylus and pteranodon and other pterosaurs. SO, there really were flying reptiles at the time of the dinosaurs, they just had a slightly different name than what we typically call them.

OH. O KAY.

PTEROSAUR FACTS!

① Pterosaurs first appeared in the Triassic period, and continued to evolve during the Jurassic period. They had wings made out of stretched skin and muscles and tissue. They came in all shapes and sizes!

← BONY CREST

② And some were HUGE!

QUETZALCOATLUS
(KWET-sal-co-AT-lus)

are thought to be some of the largest known flying animals...ever!

AS TALL AS A GIRAFFE →

(80–66 MILLION YEARS AGO)

45-FOOT WINGSPAN!

HEY! GIMME A RIDE!

BRAVE KID

SOME, LIKE THE
ANUROGNATHID,
WERE AS SMALL AS A BUTTERFLY.

(a-NEAR-og-NETH-id)

WHEEEEE!

SHORT TAIL

POSSIBLY NOCTURNAL

③ While birds and other flying animals tend to walk on TWO feet, fossilized footprints indicate that pterosaurs might have been walking around on all four feet! That means they'd have to fold their wings up when they walked.

LITTLE TAIL

④ Pretty crazy name, right? The quetzalcoatlus was named after the Aztec god of wind, air, and learning. His name was Quetzalcoatl, which means "feathered serpent."

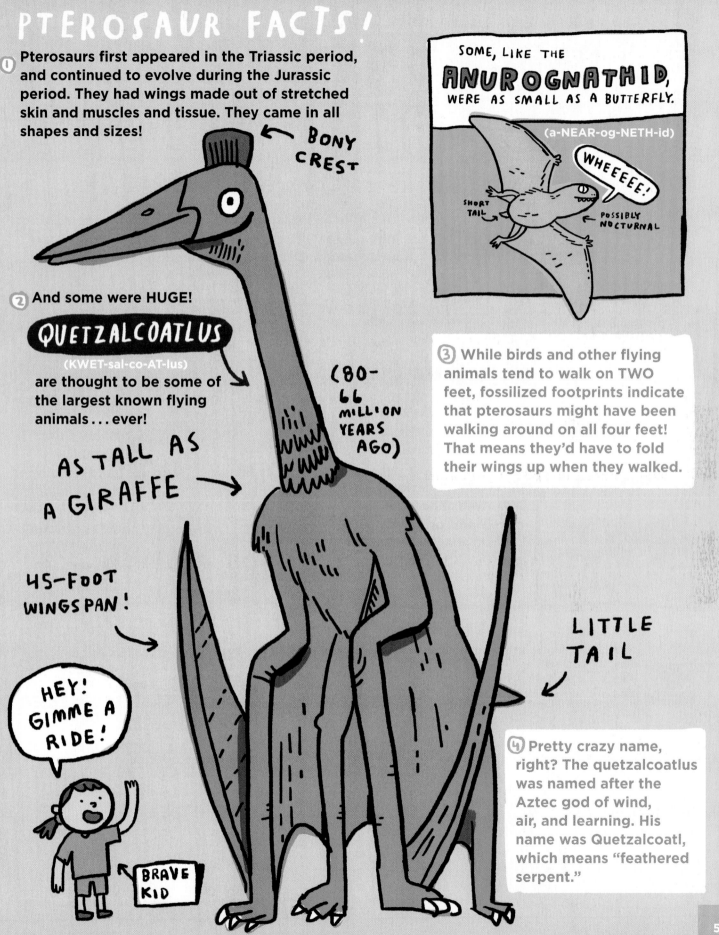

ICHTHYORNIS

SEA BIRD WITH TEETH!

95–83 MILLION YEARS AGO

(ICK-thee-OR-niss)

THE EVOLUTION of BIRD FLIGHT

(TWO THEORIES)

Bird flight is an incredibly complex form of movement for an animal. Most scientists agree that birds evolved from small dinosaurs, but how flight began is still debated.

TREE DOWN THEORY

The ancestors of birds developed the ability to glide down from trees and later developed other skills that allowed them to actually fly.

WHEEE!

GROUND UP THEORY

Birds' ancestors were small and fast and developed feathers for reasons that didn't involve flight. They later developed lift and then eventually took flight.

OH NO!

At the end of the Cretaceous period (which ended the Mesozoic era), there was ANOTHER

mass extinction.

This is the big one. It's the one that sadly killed off the dinosaurs. Read more about this event in the "Where Did They Go?" section.

PART **THREE**

MEET THE DINOS!

HELLO.

It took **BILLIONS** of years for them to show up.
Now we get to meet some of our favorite giant
reptiles that roamed the Earth millions of years ago!

DINO DINNER

WHAT DID THE DINOSAURS EAT?

MENU
① MEAT
② VEGGIES

SOME DINOS ATE MEAT (CARNIVORES) AND SOME DIDN'T EAT MEAT (HERBIVORES).

SOME ATE BOTH. → OMNIVORES

IF A DINO ONLY ATE PIZZA IT WOULD BE CALLED AN

OMNOM - NIVORE!

MOST DINOSAURS WERE HERBIVORES

They would eat leaves, small sticks, and seeds. Herbivore teeth weren't very sharp, and they were usually straight and close together, like cow's teeth. They used their teeth to rake the leaves and bark into their mouths.

CARNIVORES

Dinosaurs that ate meat had long and sharp teeth that were serrated (which means they had edges like this). → That helped them cut through the thick skin of their prey, including lizards, insects, mammals, and other dinosaurs.

And it wasn't always a battle to the death!

GASTROLITHS

Some dinosaurs would eat rocks that would end up in their gizzards. These rocks would help break down some of the stuff they ate to make it easier to digest.

Scientists believe that the fearsome T. rex was actually a scavenger and would eat animals it would find that were already dead.

ANYBODY GONNA EAT THIS?

DINOSAUR TRAITS

DID DINOSAURS

You might not like the answer to this one. We love to imagine a mighty T. rex standing over the top of a recent kill, making a loud roar to let everyone know how awesome it was. Well, we've already learned that it was probably a scavenger... and now to take T. rex down yet another step. There's a chance... that it

HONKED LIKE A GOOSE!

YOU'RE ALLOWED TO FEEL A BIT DISAPPOINTED BY THIS FACT.

Grrrrrrr

There's also a possibility that it made a low, growly, gurgling sound like an alligator.

To figure out what they sounded like, we have to look at alligators, crocodiles, and birds, which are the closest living relatives of the dinosaurs.

Birds make sound using an organ in their throat called a

SYRINX

Crocodiles and alligators make sound using a structure in their throat called a

LARYNX

(We've got these, too. Sometimes we call it our "voice box.")

RANDOM DINO FACT!

Canada has a few coins that have dinosaurs on them. One of them even

GLOWS IN THE DARK!

It features the *Pachyrhinosaurus lakustai*, which was discovered in Alberta, Canada.

CANADA
25 CENTS

DID DINOSAURS HAVE FEATHERS?

YES! WE KNOW AT LEAST SOME DINOSAURS HAD THEM!

Even if they had feathers, most dinosaurs probably couldn't fly... and maybe some could fly a little, but nothing compared to today's birds. The feathers were probably more for things like insulation or to attract a mate!

In 1966 a dinosaur was found in China that was covered in FUZZ. (Well, scientists call it "protofeathers.") It's called sinosauropteryx and it was so well preserved we can even tell that it had orange and white coloring!

Since it was found in China, here's its name in Chinese: 中华龙鸟

BUT MOST IMPORTANTLY → **WHAT ABOUT T. REX?**

Did T. rex have feathers?

Probably not. It might have had some areas of fuzz, but scientists believe that the mighty tyrannosaurus was covered in scales, not feathers.

However, we do know for sure that a cousin of the T. rex for sure had feathers.

← **YUTYRANNUS** (YOO-tie-RAN-us)

It's the largest known dinosaur to be found with preserved evidence of feathers. It's estimated that the yutyrannuses were 30 feet long and weighed one and half tons!

Have you ever seen those drawings of the T. rex battling a stegosaurus? That would've been pretty cool (and terrifying), but it definitely never happened. Some dinosaurs existed at the same time, but some were separated by MILLIONS of years.

Nope. Dinosaurs were extinct for more than
64 MILLION YEARS
before humans evolved.

GIANT DINOS!

DIPLODOCUS
(DIP-lo-DOCK-us)

For a long time it was considered the longest dinosaur. It was roughly the size of four elephants. Turn the page to see a dinosaur that's MUCH longer!

CAMARASAURUS
(CAM-a-ra-SORE-us)

TINY HEADS

APATOSAURUS
(ah-PAT-oh-SORE-us)

THEY'RE **NOT** MEAT EATERS, RIGHT?!

HUMAN ← THIS REALLY IS HOW SMALL YOU WOULD BE!

HOW DID THEY GET SO BIG?

One reason is that they didn't chew their food! Yep, they had teeth, but mos used them to rake leaves and needles from trees. Mammals don't get as big the sauropods, in part, because chewing requires a lot of energy.

THE SAUROPODS

These absolutely HUGE plant-eating dinosaurs first appeared in the late Triassic period and walked the Earth for 140 million years. They had tiny heads, big bodies, and long necks that they used to get food from tall trees. Some had loooong tails that they may have been able to crack like a whip. Some even had a club at the end of their tails.

WHOOPS!

Have you ever heard of a brontosaurus? Well, some scientists believe there's no such thing! Two years after the apatosaurus was discovered, the same paleontologist who discovered it found even larger bones of a yet-to-be-discovered new dinosaur and he named it brontosaurus. Turns out, the bones might have just belonged to an ADULT apatosaurus. However, some scientists believe that the bones belong to a different and larger species.

HUGE BODIES

LONG TAILS

BUT WHO WAS THE BIGGEST DINOSAUR?!

THE **DINO** AWARDS

BIGGEST DINO AWARD
TITANOSAURS!
(tahy-tan-uh-sawr)

Titanosaurs included some supersized sauropods! By the end of the Cretaceous period, the titanosaurs were on every continent.

ARGENTINOSAURUS
(ar-jen-TEEN-oh-SORE-us)
HERBIVORE

They could grow up to 121 feet long and weigh over 70 tons (140,000 pounds.) That's TEN TIMES the weight of the biggest elephants on Earth today and longer than a 737 plane.

SMARTEST DINO AWARD

TROODON

(TROH-oh-don)

IT HAD AN UNUSUALLY LARGE BRAIN

CARNIVORE

that might have made it a little smarter than other dinosaurs, but still only about as smart as a chicken.

WHAT'S THAT SUPPOSED TO MEAN?!

Troodons had eyes set toward the front of their faces, rather than on the sides of their heads. This gave them the ability to be better hunters.

Many herbivores had eyes on the sides of their heads so they could be better at noticing if a predator was approaching.

HI

WEIRD FACT →

One scientist speculated that if the troodons weren't killed off in the mass extinction, they possibly could have evolved into intelligent creatures with bodies similar to humans... but of course this didn't really happen!

THE, UM, NOT-SO-SMARTEST DINOSAUR AWARD GOES TO:

STEGOSAURUS
(steg-oh-SORE-us)

Okay, they probably weren't the dumbest dinosaurs, but they did have unusually small brains!

BRAIN THE SIZE OF A LIME!

HERBIVORE

THEIR SPIKED TAIL IS CALLED A **THAGOMIZER!**

Compared to the size of their bodies, dinosaurs had much smaller brains than mammals do. In humans, most of what is under the surface of the skull bone is brain matter, but in a dinosaur a lot of that space was taken up with a giant jawbone and strong muscles used for biting.

PALEONTOLOGISTS FOUND IT SO HARD TO BELIEVE THAT THE STEGOSAURUS HAD SUCH A SMALL BRAIN THAT FOR A WHILE SOME BELIEVED THEY HAD EXTRA BRAIN MATTER IN THEIR HIP REGION!

(IT TURNED OUT TO JUST BE A POCKET OF EXTRA TISSUE.)

TRICERATOPS
(try-SERR-ah-tops)

MEANS "HORN FACED" (FOR REAL)

THEY HAD REALLY BIG HEADS!

← 30 FEET LONG

BEAK LIKE A PARROT

CAN IT TALK?

THEY WOULD CHARGE LIKE A RHINO!

HERBIVORE

Sure, the horns on these dinosaurs look...um...a little intimidating, but some scientists believe they might have been for attracting a mate, not for fighting off enemies. Whatever the reason, these dinosaurs had a bone that stuck out of their heads that was covered in a material called keratin, which was a lot like human fingernails!

Triceratops weren't the only dinosaurs with awesome horns. In fact, dozens of dinosaurs had them! Here are some of my favorites:

HERBIVORE

FOUND IN CANADA

THESE BUMPS ARE CALLED "BOSSES"

WEIRD FACT! THEY STARTED MATING AT NINE YEARS OLD!

(pack-ee-RHINE-oh-SORE-us)

STYRACOSAURUS (stih-RAK-uh-SAWR-us) (GREEK FOR "SPIKED LIZARD")

PACHYRHINOSAURUS (GREEK FOR "THICK-NOSED LIZARD")

→ MAY HAVE BEEN FASTER THAN AN ELEPHANT! (ABOUT 15 MILES PER HOUR)

BUT WHO HAD THE MOST HORNS?!

15 HORNS!

HERBIVORE

MOST HORNS AWARD

KOSMOCERATOPS!

(KOS-mo-SERR-a-tops)

NAME MEANS "ORNATE HORN FACE"

DEVELOPED ALL THIS CRAZY ORNAMENTATION TO ATTRACT A MATE

SUPER-FAST DINOSAURS

OUTTA MY WAY!

THEY WERE ABOUT 12 FEET LONG.

200 POUNDS

LARGE EYES

OMNIVORE

DROMICEIOMIMUS
(dro-MISS-ee-oh-mee-mus)

THE ORNITHOMIMIDS

THEY'RE A GROUP OF DINOSAURS THAT LOOKED LIKE OSTRICHES BECAUSE OF THEIR SIZE AND BIG LEGS!

I CALL THEM "FAST FOOD"!

THEY COULD RUN 50 MPH!

WEIRDEST LOOKIN' DINO

(Scientists have nicknamed them ELVISAURUSES because of that crazy crest thing on top of their heads.)

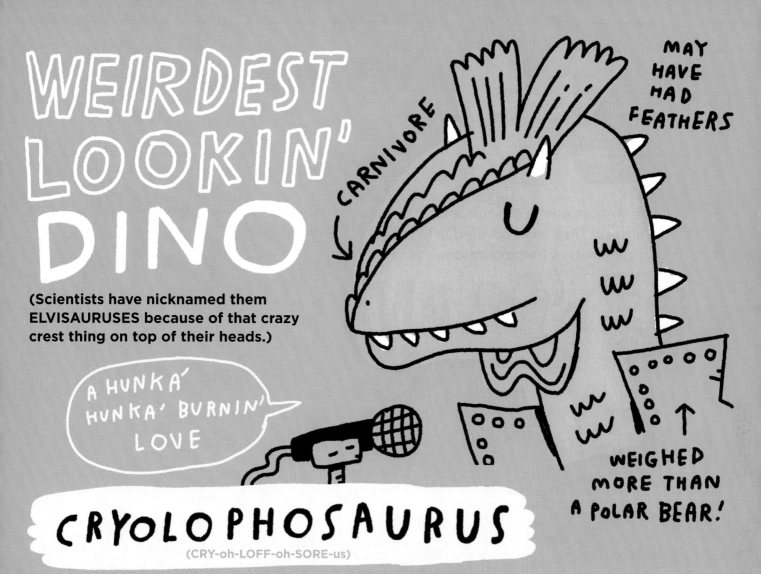

CARNIVORE

MAY HAVE HAD FEATHERS

WEIGHED MORE THAN A POLAR BEAR!

A HUNKA' HUNKA' BURNIN' LOVE

CRYOLOPHOSAURUS
(CRY-oh-LOFF-oh-SORE-us)

When you think of Antarctica, you imagine it covered in ice, right? Well, during the early Jurassic period, when the cryolophosaurus was around, Antarctica was much closer to the equator and NOT frozen at all! In fact, the entire planet was warmer and there were no ice caps at all at the North and South Poles.

JOKE TIME!

I DON'T GET IT.

Scientists can tell it was warm because there were no scarves or mittens found near the cryolophosaurus's bones.

ARMORED DINOSAURS!

I CAN'T SEE!

The ankylosaurs and nodosaurs were huge, plant-eating dinosaurs that lived during the Mesozoic era. They were covered in hard plates and spiky armor that they would use to protect themselves from predators.

ANKYLOSAURUS

THEY WERE BUILT LIKE PREHISTORIC TANKS!

SOME HAD CLUBBED TAILS THAT COULD BREAK A T. REX'S BONES!

6 FT TALL

5-6.5 TONS

SMALL BRAINS

ATE LOW-LYING PLANTS

(an-KILE-oh-SORE-us)

THEIR HEAD, NECK, AND TAIL WERE COVERED WITH HARD PLATES OF BONE CALLED OSTEODERMS.

WANNA RACE?

They were so heavy and low to the ground, they could only go about 6 miles per hour!

ANTARCTOPELTA

(ann-TARK-toh-PEL-ta)

13 FEET LONG

FOUND IN ANTARCTICA (MAYBE YOU COULD GUESS THAT?)

MIDDLE CRETACEOUS

LOOKS LIKE IT SHOULD BREATHE FIRE, BUT IT DOESN'T.

EDMONTONIA

(ed-mon-TONE-ee-ah)

CHECK OUT THAT HELMET

DIDN'T GET A CLUBBED TAIL! (AWW)

LATE CRETACEOUS

THIS LITTLE BUG HAS A QUESTION.

HOW OLD WOULD THE DINOSAURS GET?

Scientists once believed that dinosaurs could live for hundreds of years because of how massive they grew. More recently, we have learned that they could grow very quickly. Carnivores are believed to have lived to be around 30 years old, while herbivores like the giant sauropods could live up to 100 years.

WHAT'S THAT? SPEAK UP!

HERE ARE TWO COUSINS THAT (BOTH) GET AWARDS.

LONGEST NAME AWARD!

MICRO-
PACHY-
CEPHALO-
SAURUS

WHICH MEANS "SMALL, THICK-HEADED LIZARD"

THAT'S NOT VERY FLATTERING!

← 5-10 POUNDS

2 FEET LONG

(MY-crow-PACK-ee-KEFF-a-lo-SORE-us)

SO THE T. REX WASN'T THE BIGGEST OR THE FASTEST DINOSAUR BUT IT DID HAVE REALLY BIG TEETH!

They could use their super-strong jaws to rip off over 500 pounds of meat from their prey with a single bite.

BIGGEST TEETH AWARD!

ITS TEETH WERE THE SIZE OF BANANAS!

THAT'S BANANAS.

BUT, UM, MUCH SHARPER.

Their arms were only 3 feet long. That might seem little, but they weren't useless! It's believed that they were used to hold prey while eating or to SLASH prey if they got too close.

TYRANNOSAURUS REX!

(tie-RAN-oh-SORE-us)

NAME MEANS:
"TYRANT LIZARD KING"

- COULD GROW 15-20 FEET TALL
- THEY HAD BIG BRAINS! (ABOUT TWICE AS BIG AS OTHER PREDATORS THEIR SIZE)

They grew VERY QUICKLY. Between the ages of 14 and 18, they would put on an average of 5 pounds every single day.

More than 20 almost-complete T. REX skeletons have been found. The most intact one is named Sue.

CONTRARY TO POPULAR BELIEF, the T. REX actually had really great eyesight (and they were fast enough to have caught humans if we had lived in the time of the dinosaurs)!

DINOSAUR EGGS

Dinosaurs laid eggs just like birds do today; however their eggs had really thick shells.

mommy?

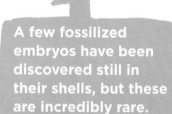

A few fossilized embryos have been discovered still in their shells, but these are incredibly rare.

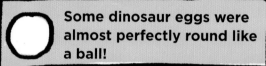

When we think of an egg, it's shaped like this.

Some dinosaur eggs were almost perfectly round like a ball!

Some were elongated like a loaf of bread.

WEIRD FACT!

SOME DINOSAURS MAY HAVE HAD

OUCHIE.

DINO-SIZED FLEAS!

YUM!

HAD CLAWS THAT HELPED THEM HOLD ONTO THEIR PREY.

SUPER-SHARP, SERRATED TUBES COULD PIERCE THICK DINO SKIN.

THEY WERE 10 TIMES THE SIZE OF MODERN-DAY FLEAS.

PART FOUR

WHERE DID THEY GO?

EXTINCTION

65 MILLION YEARS AGO, SOMETHING KILLED THE DINOSAURS.

(AND THE PTEROSAURS, THE GIANT MARINE REPTILES, AND ABOUT 75% OF ALL SPECIES ON EARTH.)

DIED

DIED

IT'S CALLED THE CRETACEOUS-PALEOGENE EXTINCTION EVENT.

(AKA THE K-Pg EXTINCTION)

So what happened? Well, we don't know for sure, but scientists have a few theories.

SORRY.

COUGH, COUGH!

THEORY ①

One theory is that volcanoes are to blame for their extinction. There was a huge increase in volcanic eruptions and all of that gas in the atmosphere could have trapped a lot of heat . . . making it too hot for the dinosaurs. OR, all of the volcanic eruptions caused ash and dust to fill the atmosphere, blocking out the sun.

THEORY 2

Another theory is that an asteroid 6 to 10 miles wide hit the Earth. The impact would have caused dense clouds of dust all over the world to block out the sun. This would have killed off the plants, which would then kill off plant-eating dinosaurs and then dinosaur-eating dinosaurs.

OUCHIE!

CANADA

USA

MEXICO

There's a huge crater in an area in Mexico called

CHICXULUB

that might be where the asteroid hit the Earth. (It's about 110 miles across! That means the crater is roughly the same width as Connecticut!)

THE IMPACT WOULD HAVE BEEN 2 MILLION TIMES MORE POWERFUL THAN AN ATOMIC BOMB!

YIKES!

CHICXULUB CRATER

A FEW OTHER THEORIES:

① AN INCREASE IN DINO EGG-EATING MAMMALS

② PERIODS OF EXTREME COLD

③ PERIODS OF EXTREME HEAT

④ WIDESPREAD DISEASE

Or, it could have been a combination of a few of these theories! Whatever it was, it wasn't an overnight event. The extinction of the dinosaurs took place over thousands of years!

The K-Pg extinction killed off the dinosaurs and ended the Mesozoic era, but it was only the THIRD-largest mass extinction event.

A FEW KINDA WEIRD DINO EXTINCTION THEORIES

AND UNLIKELY! ←

NOT ENOUGH FEMALES.

Did climate change lead to only male dinosaurs, which made it impossible to reproduce? (Nope.)

ALIENS.

It would be cool if aliens caused extinction. (But they didn't.)

BUGS!

Maybe too many caterpillars ate all the dino food? (Probably not.)

AND THE AWARD FOR THE BEST ALL-TIME SUPER-WEIRD DINO EXTINCTION THEORY GOES TO:

DINOSAUR FARTS

Okay, so this wasn't ever really a scientific theory. It was a misunderstanding after some scientists tried to calculate how much methane gas the dinosaurs may have generated to see if it could have influenced the climate. Some news sources read about the research and reported that it was a new theory that dinosaur farts are the reason why the dinosaurs went extinct.

EXCUSE ME.

DID YOU KNOW → THE PHRASE: "GOING THE WAY OF THE DINOSAUR" IS AN EXPRESSION USED FOR SOMETHING THAT USED TO BE COMMON BUT ISN'T ANYMORE. (CAN YOU THINK OF ANY EXAMPLES?)

EXTRA! EXTRA!

NOT ALL THE DINOSAURS DIED!

RAWR!

HAVE YOU EVER HEARD THAT **BIRDS** EVOLVED FROM **DINOSAURS??**

WELL, THEY **ARE** DINOSAURS!

Birds evolved from dinosaurs during the Jurassic period, roughly 150 million years ago. While some died out during the mass exctinction that killed the dinosaurs, some survived.

And they still walk... er, um... FLY among us today!

AND HERE'S MY GREAT-GREAT GRANDMA...

TOTALLY WEIRD BONUS FACT

There's a small town called Dinosaur, Colorado (USA). Some of its street names include Brontosaurus Blvd., Brontosaurus Bypass, Stegosaurus Freeway, and Tyrannosaurus Trail.

Welcome to **Dinosaur, Colorado** GATEWAY TO DINOSAUR NATIONAL MONUMENT

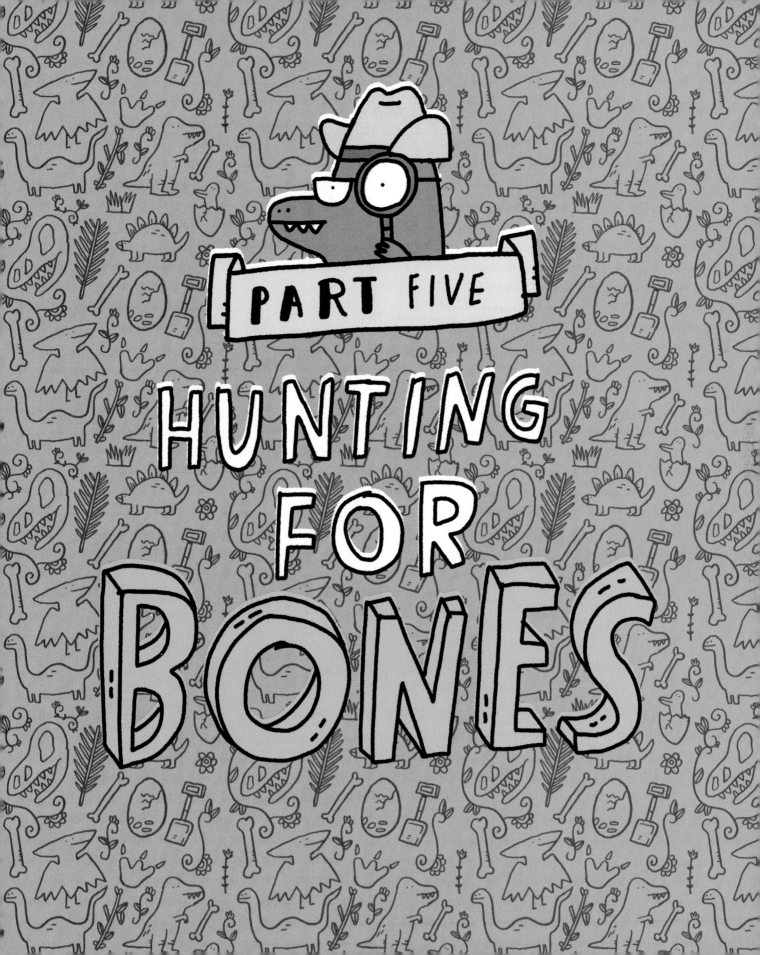

PART FIVE

HUNTING FOR BONES

IF DINOSAURS DIED OFF MILLIONS OF YEARS AGO, HOW CAN WE KNOW SO MUCH ABOUT THEM?

When the dinosaurs died, they left clues that can help us figure out what they looked like, how big they were, what their environment was like, what they ate, and sometimes even stuff about their personality.

LIKE WHAT MY FAVORITE COLOR IS?

NOT EXACTLY.

More like if they traveled in big groups called herds (like the giant sauropod the camarasaurus, who would travel for 200 miles to and from the plains to the mountains to find food and water).

♫ ON THE ROOAAD AGAAAIN ... ♫

CAMARASAURUS HAD STIFF NECKS.

THEIR FOSSILS ARE SUPER COMMON.

COOL SIDE FACT! →

We are currently in a "golden age" of dinosaur discovery, with a new species of dinosaur being discovered almost every week! This is possible with new technology and the fact that some countries like China, Mongolia, and Argentina have only recently allowed paleontologists to explore them. Almost HALF of all the new dinosaurs discovered have been in China.

WHAT?!

DINOSAURS LEFT BEHIND
FOSSILS.

STUFF LIKE:

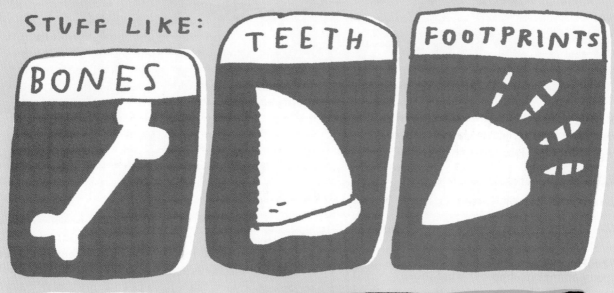

THEY EVEN LEFT BEHIND FOSSILIZED... UM...DROPPINGS!

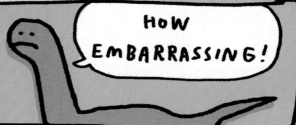

HOW EMBARRASSING!

These remains have been found all over the world for centuries and are probably how we ended up with stories about dragons and some other mythical creatures.

PALEONTOLOGISTS

ARE SCIENTISTS WHO STUDY

FOSSILS.

They hunt for bones on expeditions where they spend long days searching for signs left by the dinosaurs.

They also compare the fossils they find to living animals to find out what those dinosaurs could have been like. If a skeleton shows large, sharp teeth and thin limbs, we can assume that it was a carnivore that needed to move quickly to catch its prey.

THAT'S RIGHT.

Piecing together all this stuff can be like putting together a really big puzzle with missing pieces and no picture to look at. In fact, there have been some pretty crazy assumptions made by scientists in the past. Not because they weren't smart, but because they didn't have all the information we have today.

AMBER

Scientists have even found a piece of a dinosaur tail trapped in ancient, hardened tree sap called amber. Are you wondering if the piece of dinosaur tail had feathers? It did!

BONUS
- - - - -
FACT!

One of the biggest bones ever found is the backbone of the argentinosaurus. It was 5 ft. by 5 ft. and weighed more than a TON.

THE CONDITIONS HAVE TO BE JUST RIGHT TO MAKE A FOSSIL.

HERE'S AN EXAMPLE:

The dinosaur dies.

Its flesh is eaten by other animals.

ROTTING FLESH! YUM!

The bones sink into the mud.

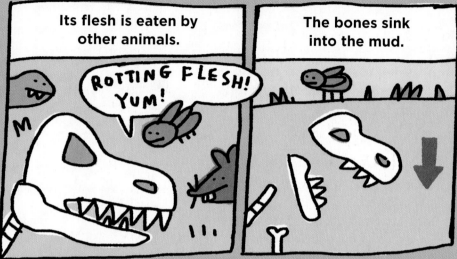

Over millions of years, layers of rock and sand settle on top of the bones.

Some of those layers of sediment get worn away by water or wind, until the bones are close enough to the surface for them to be seen.

A SKULL!

And, the bigger the dinosaur, the harder it is to find a complete skeleton. Bones can get separated, or break apart, or get so buried that no one ever finds them. But smaller skeletons can get covered by mud and dirt more quickly, which helps to keep the bones together.

A TYRANNOSAURUS REX NAMED SUE

The largest, most complete T. rex ever discovered was in 1990 by Sue Hendrickson in South Dakota. The fossil is very well preserved, and is almost 90% complete. In 1997, Sue was sold for $8.3 million, the highest amount ever paid for a dinosaur fossil. Its permanent home is the Field Museum of Natural History in Chicago.

BATTLE BONES!

SO COOL!

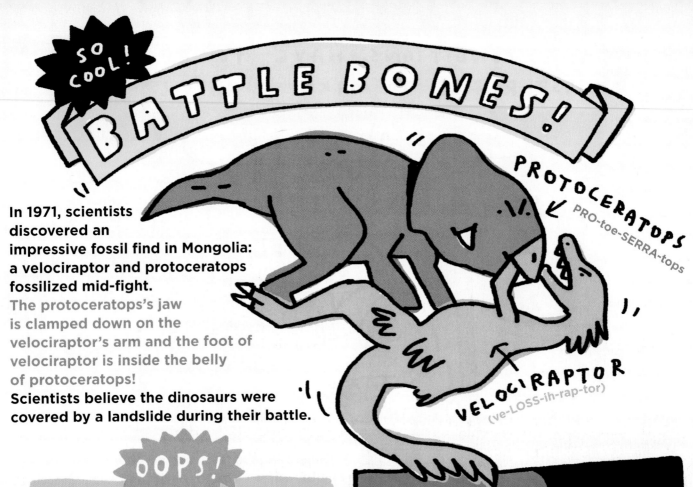

PROTOCERATOPS
(PRO-toe-SERRA-tops)

VELOCIRAPTOR
(ve-LOSS-ih-rap-tor)

In 1971, scientists discovered an impressive fossil find in Mongolia: a velociraptor and protoceratops fossilized mid-fight.
The protoceratops's jaw is clamped down on the velociraptor's arm and the foot of velociraptor is inside the belly of protoceratops!
Scientists believe the dinosaurs were covered by a landslide during their battle.

OOPS!

Fossils are everywhere ... some of the best places to find them are river valleys, cliffs, and hillsides. And many fossils are found by complete accident.

In fact, a nine-year-old boy discovered a million-year-old fossil when he tripped over it while hiking with his family. Jude Sparks was hiking in the mountains of New Mexico when he tripped over an unusual bone sticking out of the ground. Turns out, it was a stegomastodon fossil, a distant cousin to ancient mammoths and modern elephants. It's one of only 200 stegomastodon fossil sets ever found.

STEGOMASTODON
(STA-go-mas-ta-DON)

LA BREA TAR PITS

In Los Angeles there's an area where tar bubbles right out of the ground. For centuries, the super-sticky stuff has preserved the remains of the animals that have gotten stuck there. They've found saber-toothed cats, mammoths, AND EVEN A HUMAN BODY!

PART SIX

MORE PREHISTORIC BEASTS!

When the dinosaurs died, the Mesozoic era ended and the Cenozoic era began. It's the era we are currently in now! It started 66 million years ago, and there have been some pretty cool animals that have lived (and died) since then.

TOP 13 MOST AWESOME EXTINCT CENOZOIC BEASTS

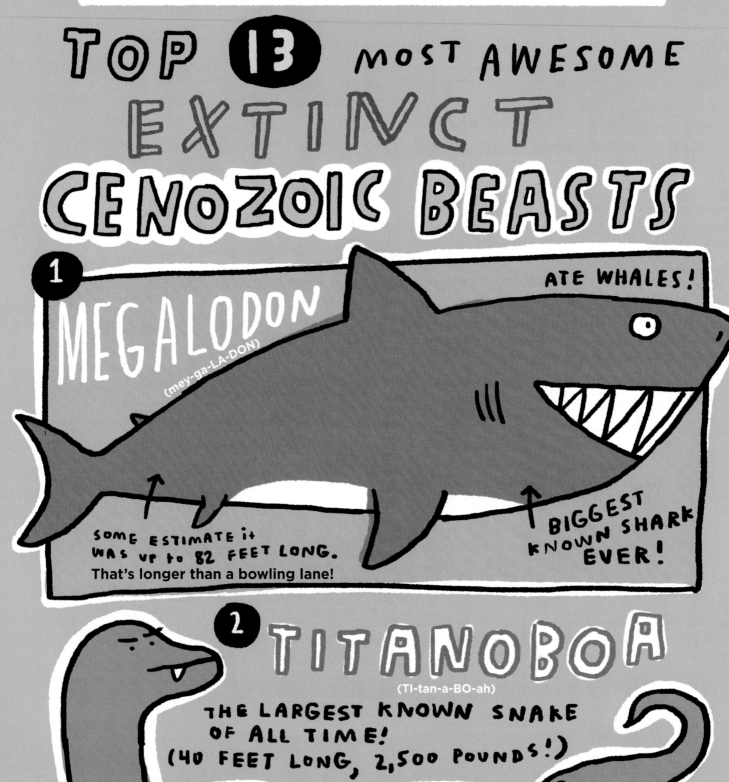

1 MEGALODON (mey-ga-LA-DON)

ATE WHALES!

SOME ESTIMATE it WAS UP to 82 FEET LONG. That's longer than a bowling lane!

BIGGEST KNOWN SHARK EVER!

2 TITANOBOA (TI-tan-a-BO-ah)

THE LARGEST KNOWN SNAKE OF ALL TIME! (40 FEET LONG, 2,500 POUNDS!)

WOOLLY MAMMOTHS ③

(WULL-ee MAMM-oth)

In 2012, an 11-year-old boy in Russia happened upon an extremely well-preserved woolly mammoth carcass while walking his dogs.

They were roughly about the size of modern African elephants.

Scientists can figure out a woolly mammoth's age from the rings of its tusk, like looking at the rings of a tree.

Their large, curved tusks may have been used for fighting. They also may have been used as a digging tool for foraging meals of shrubs, grasses, roots, and other small plants from under the snow.

The woolly mammoth was not the only "woolly" type of animal. Here's a woolly rhinoceros, also known as *coelodonta antiquitatis*.

4

CRAZY FACT!

WOOLLY MAMMOTHS DIED OUT ONLY 4,000 YEARS AGO. THAT MEANS THEY WERE STILL ALIVE WHEN THE PYRAMIDS WERE BUILT!

5 ENTELODON
(en-TELL-oh-don)

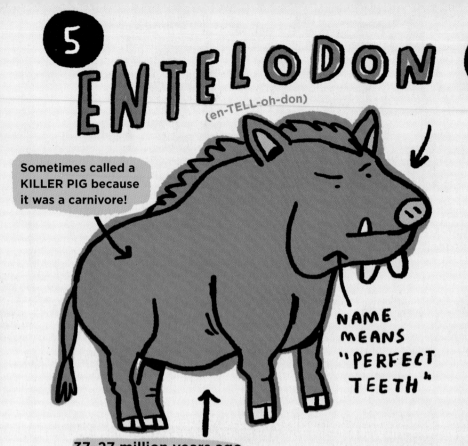

Sometimes called a KILLER PIG because it was a carnivore!

NAME MEANS "PERFECT TEETH"

37–27 million years ago

6 ELASMOTHERIUM
(ell-AZZ-moe-THEER-ee-um)

This furry, mammoth-sized rhino looks like a unicorn, but sadly unicorns are just folklore. (Sorry!)

7 DEINOTHERIUM
(DIE-no-THEER-ee-um)

One of the largest mammals EVER

Similar to elephants, but had two tusks attached to their lower jaw that pointed downward

ARSINOITHERIUM

(AR-sin-oh-ee-THEER-ee-um)
Lived about 27 million years ago

TWO GIANT HORNS

Looked kinda like a rhino

12

DO YOU KNOW HOW TO KEEP A RHINO FROM CHARGING?

TAKE AWAY HIS CREDIT CARD!

AND LAST, BUT NOT LEAST

13

DOEDICURUS

(dee-DI-kew-russ)

Thick, rigid armor that wouldn't flex like the armor of armadillos

This heavily armored relative of an armadillo lived in South America.

Its name means "pestle tail."

PESTLE

MORTAR

A pestle is a blunt tool used to grind stuff up, often in a bowl called a mortar.

COULD GROW UP TO 13 FEET LONG!

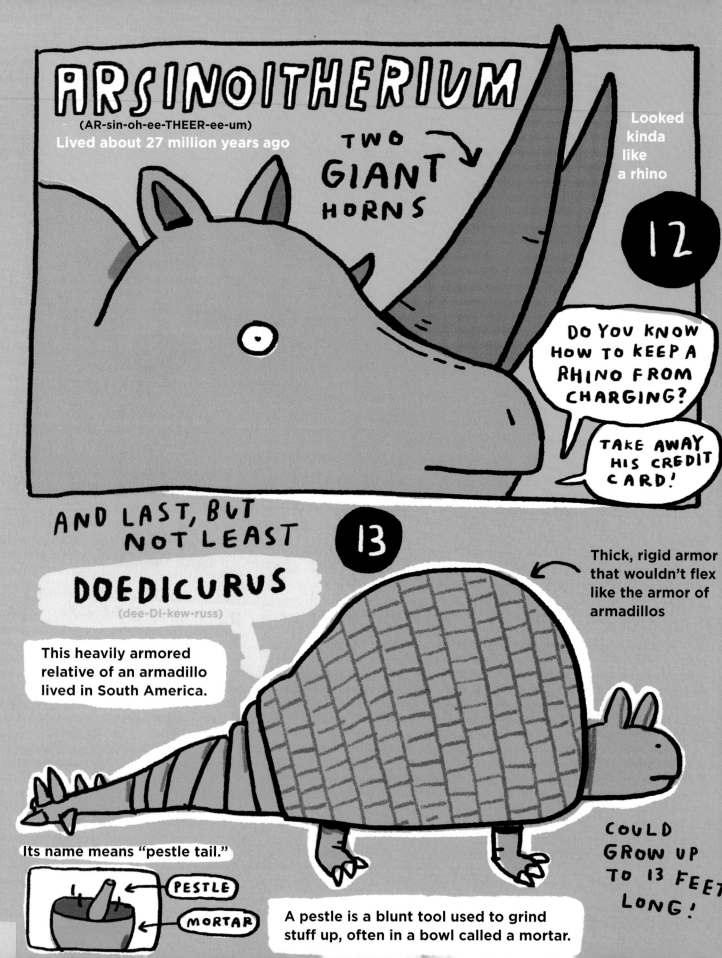

PART SEVEN

LET'S DRAW DINOSAURS!

LET'S DRAW DINOSAURS!

You don't need fancy art supplies to draw dinosaurs! Here are a few things that I like to draw with, but you can use whatever you've got.

SUPPLIES:

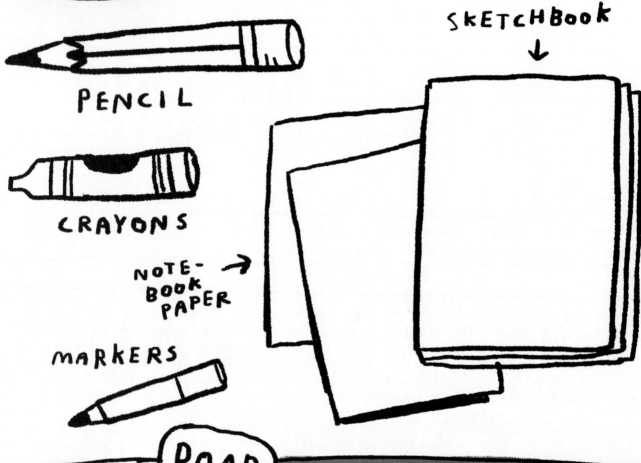

PENCIL

CRAYONS

NOTE-BOOK PAPER →

MARKERS

SKETCHBOOK ↓

ROAR

GRRR

IMPORTANT!

It's important to know that your drawings will be significantly better if you make dinosaur sounds and faces while making them. Especially if you're in a crowded, quiet area.

T. REX

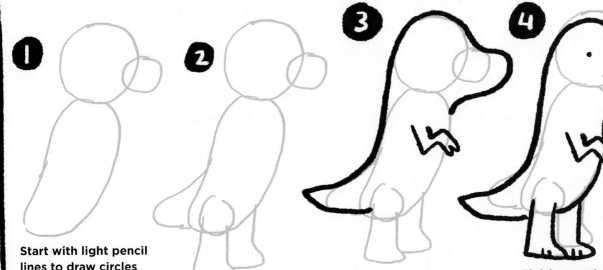

1 Start with light pencil lines to draw circles for the head, snout, and body.

2 Draw shapes for the tail and legs.

3 Use a pen or crayon to draw over your pencil lines.

4 Finish your drawing with a mouth, teeth, and tiny T. rex arms!

STEGOSAURUS

1 Start with light pencil lines to draw a big oval and a little oval.

2 Draw circles for the legs and tail.

3 Start inking your drawing with a pen and add a face.

4 Add scales, and use a pen or crayon to draw the finished stegosaurus.

BRACHIOSAURUS

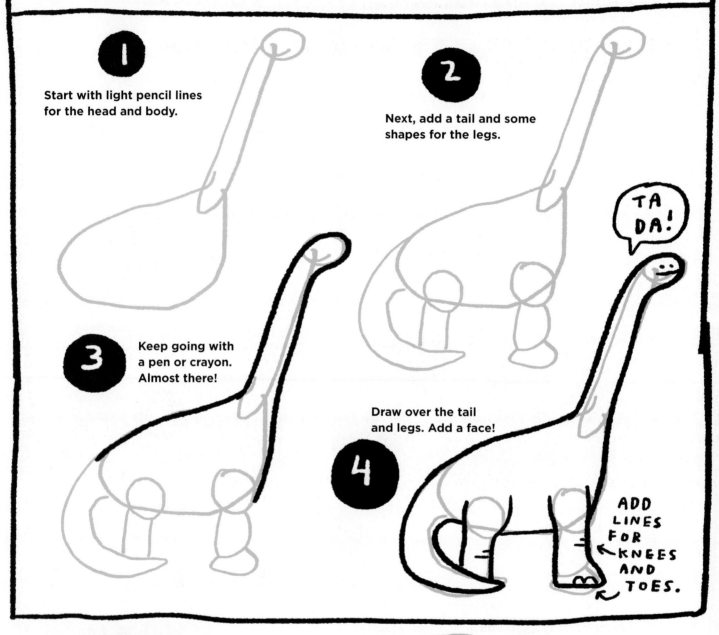

1 Start with light pencil lines for the head and body.

2 Next, add a tail and some shapes for the legs.

3 Keep going with a pen or crayon. Almost there!

4 Draw over the tail and legs. Add a face!

TA DA!

ADD LINES FOR KNEES AND TOES.

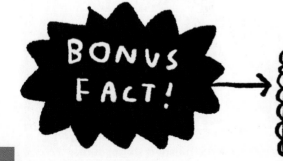

BONUS FACT!

Earth's rotation is slowing at a rate of approximately 17 milliseconds a century; that means the days are getting longer. The length of a day for the dinosaurs was closer to 22 hours.

Okay, so maybe they're not REALLY dinosaurs. Who cares!?! They're still fun to draw!

This is a tough one. Think you can handle it?

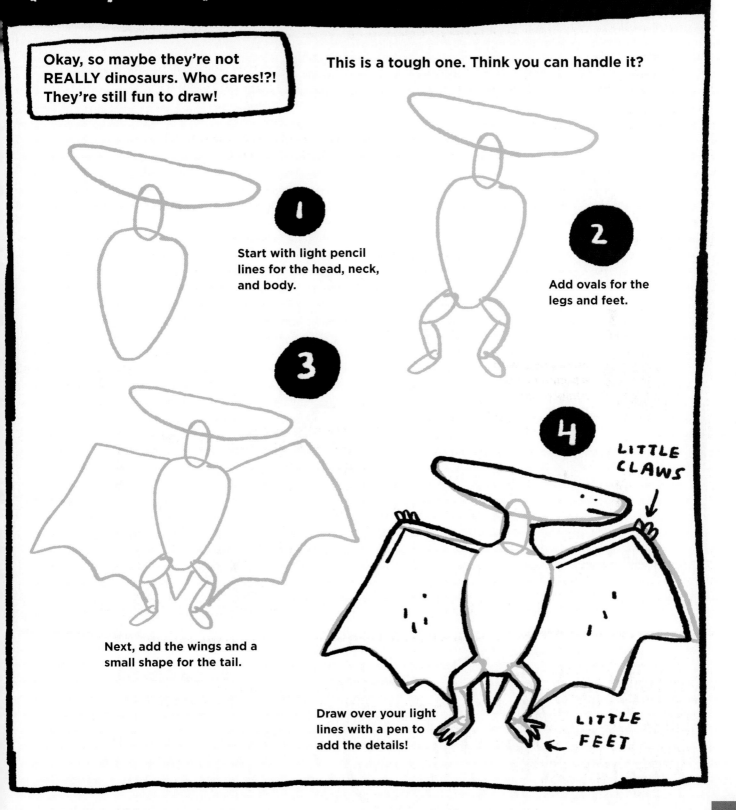

1 Start with light pencil lines for the head, neck, and body.

2 Add ovals for the legs and feet.

3 Next, add the wings and a small shape for the tail.

4 Draw over your light lines with a pen to add the details!

LITTLE CLAWS

LITTLE FEET

SPINOSAURUS

1 Start with a pencil to draw light lines. Make simple shapes like this for the head and body.

2 Draw simple shapes for the tail, sail, arms, and legs.

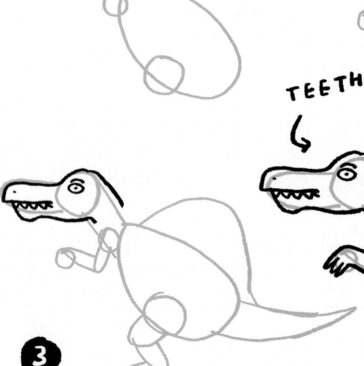

TEETH

SAIL

ADD SOME SCALES!

3 Start drawing over your pencil lines with a pen or marker. Add a face.

4 Keep drawing over your light lines, and add the details!

WEIRD BONUS FACT

SHHHHH

SOME DINOSAURS MAY HAVE SECRETLY PUT THEIR EGGS IN OTHER DINOSAURS' NESTS AS A WAY TO TRICK THEM INTO RAISING THEIR YOUNG.

Some modern birds do this today!

PART EIGHT

DINO FIELD GUIDE

A COLLECTION OF DINOSAURS LISTED BY TIME PERIOD

TRIASSIC PERIOD

251-199 MYA

DESMATOSUCHUS
(des-mat-TOE-sue-CUSS)
16 feet long • 660 lbs.

HERRERASAURUS
(her-AYR-ah-SORE-us)
10 feet long • 420 lbs.

SELLOSAURUS
(sell-OH-SORE-us)
7 feet long • 880 lbs.

THECODONTOSAURUS
(THEEK-oh-DON'T-oh-sore-us)
8 feet long • 90 lbs.

PLATEOSAURUS
(PLAT-ee-oh-SORE-us)
26 feet long • 1.5 tons

TANYSTROPHEUS
(tan-nis-TRO-fee-uss)
20 feet long • 330 lbs.

HYPERODAPEDON
(hi-per-O-dap-PEE-don)
4 feet long • 90 lbs.

CHASMATOSAURUS
(kaz-ma-ta-SORE-us)
6.5 feet long • weight unknown

STAURIKOSAURUS
(STORE-ick-oh-SORE-us)
6.5 feet long • 90 lbs.

PLACERIAS
(plah-SEE-ree-ass)
4-11 feet long • up to 1 ton

Kid for size
comparison

LAGOSUCHUS
(LAY-go-SUE-cuss)
12 inches long
(about the
size of a
ferret)

Hand for size
comparison

COELOPHYSIS
(see-LOF-ih-sis)
9 feet long • 110 lbs.

EORAPTOR
(EE-oh-rap-tor)
3 feet long • 22 lbs.

JURASSIC
PERIOD
201-145 MYA

MAMENCHISAURUS
(ma-MEN-chi-SORE-us)
72 feet long • 22 tons

CAMARASAURUS
(CAM-a-ra-SORE-us)
75 feet long • 22 tons

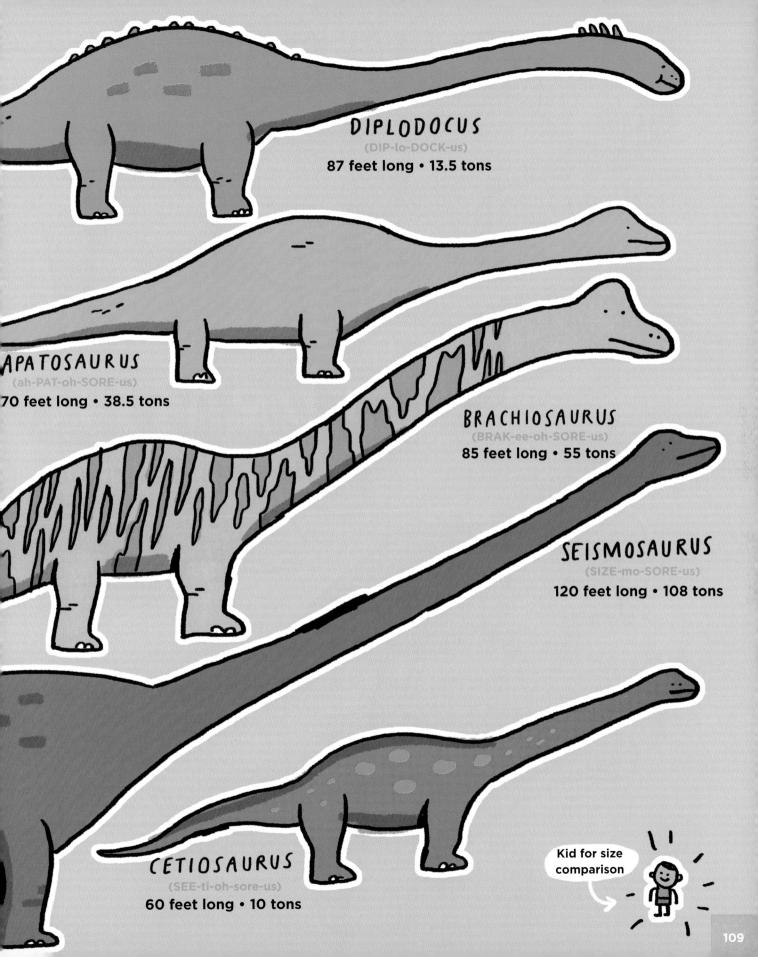

DIPLODOCUS
(DIP-lo-DOCK-us)
87 feet long • 13.5 tons

APATOSAURUS
(ah-PAT-oh-SORE-us)
70 feet long • 38.5 tons

BRACHIOSAURUS
(BRAK-ee-oh-SORE-us)
85 feet long • 55 tons

SEISMOSAURUS
(SIZE-mo-SORE-us)
120 feet long • 108 tons

CETIOSAURUS
(SEE-ti-oh-sore-us)
60 feet long • 10 tons

Kid for size comparison

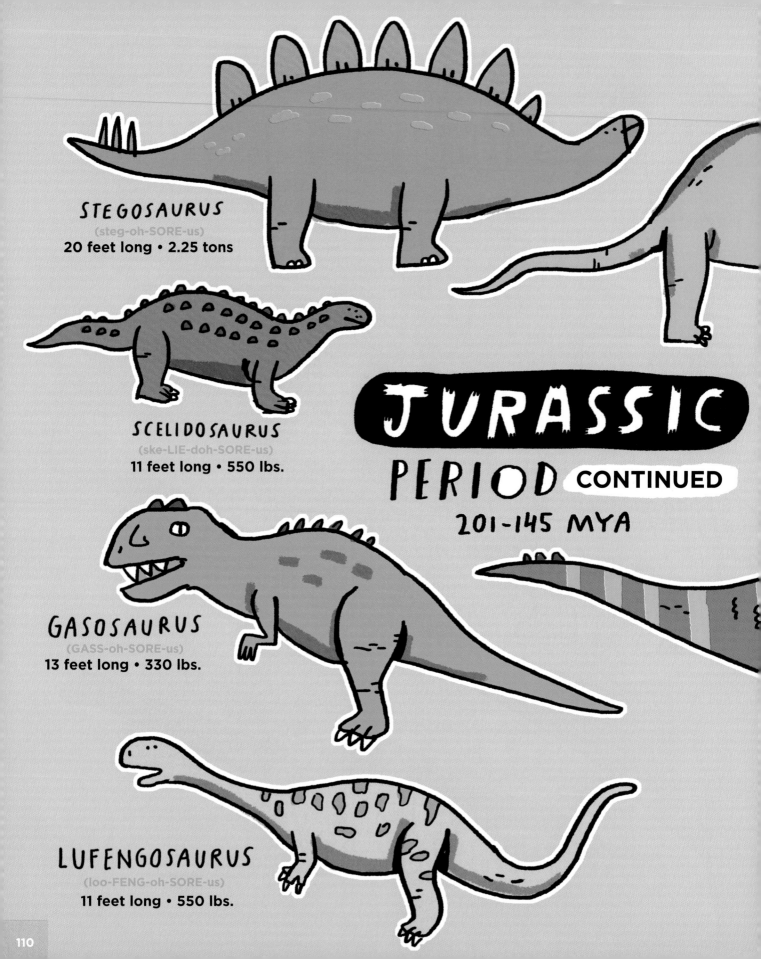

STEGOSAURUS
(steg-oh-SORE-us)
20 feet long • 2.25 tons

SCELIDOSAURUS
(ske-LIE-doh-SORE-us)
11 feet long • 550 lbs.

JURASSIC
PERIOD CONTINUED
201-145 MYA

GASOSAURUS
(GASS-oh-SORE-us)
13 feet long • 330 lbs.

LUFENGOSAURUS
(loo-FENG-oh-SORE-us)
11 feet long • 550 lbs.

VULCANODON
(vul-KAN-o-don)
21 feet long • 5 tons

GARGOYLEOSAURUS
(gar-GOY-lee-o-sore-us)
11 feet long • 550 lbs.

SHUNOSAURUS
(SHOO-no-SORE-us)
33 feet long • 11 tons

ALLOSAURUS
(AL-oh-SORE-us)
40 feet long • 3 tons

ORNITHOLESTES
(OHR-nith-oh-LEST-eez)
6.5 feet long • 26 lbs.

Kid for size comparison

ANCHISAURUS
(ankee-SORE-us)
24 inches long
6.5 lbs.

COMPSOGNATHUS
(comp-sog-NAYTH-us)
24 inches long • 6.5 lbs.

CRETACEOUS PERIOD
145-66MYA

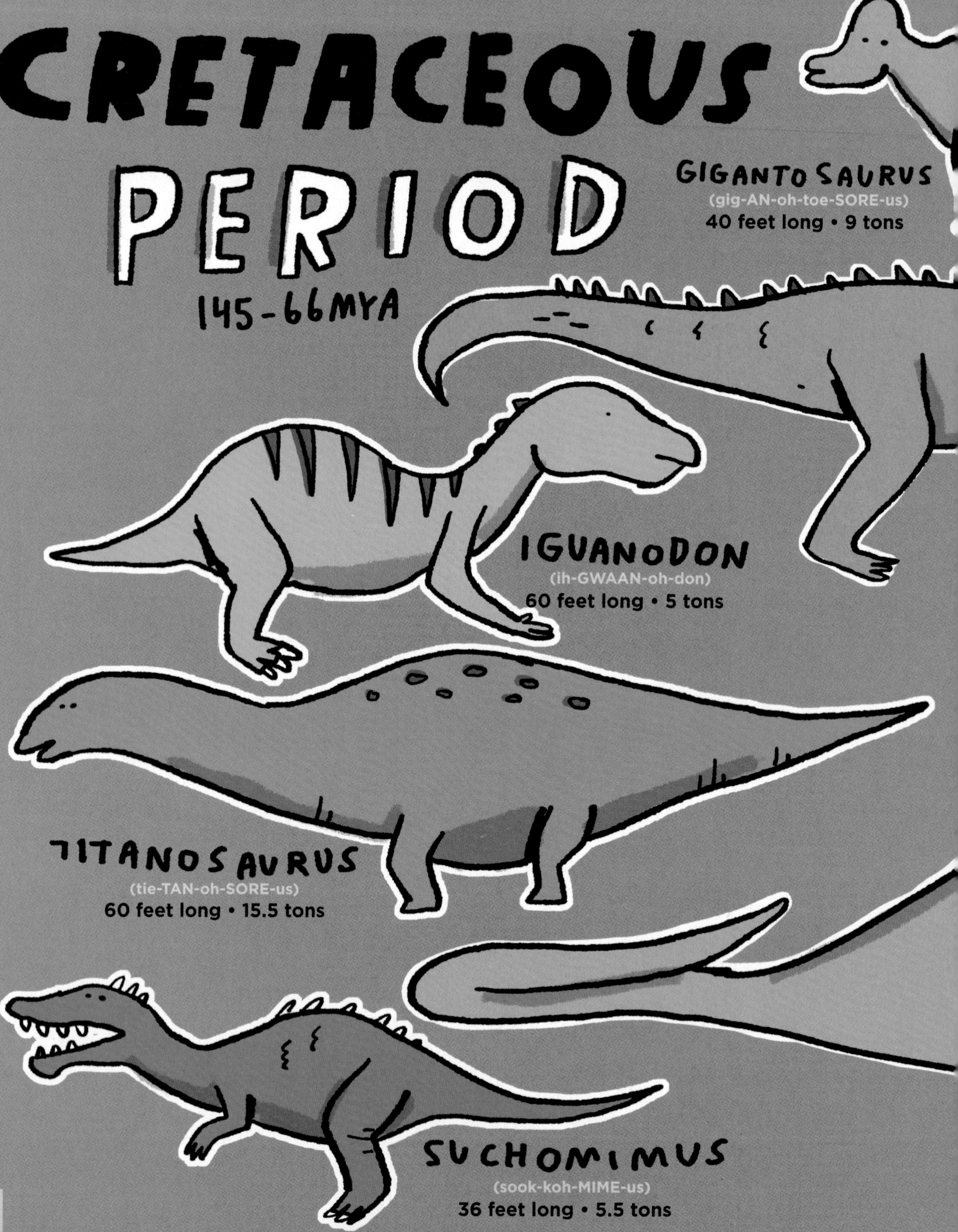

GIGANTOSAURUS
(gig-AN-oh-toe-SORE-us)
40 feet long • 9 tons

IGUANODON
(ih-GWAAN-oh-don)
60 feet long • 5 tons

TITANOSAURUS
(tie-TAN-oh-SORE-us)
60 feet long • 15.5 tons

SUCHOMIMUS
(sook-koh-MIME-us)
36 feet long • 5.5 tons

CORYTHOSAURUS
(ko-RITH-oh-SORE-us)
16 feet long • 2.5 tons

PARASAUROLOPHUS
(PAR-a-sore-OL-o-fuss)
33 feet long • 4.5 tons

SPINOSAURUS
(SPY-no-SORE-us)
50 feet long • 7.75 tons

ARGENTINOSAURUS
(ar-jen-TEEN-oh-SORE-us)
130 feet long • 110 tons

Kid for size comparison

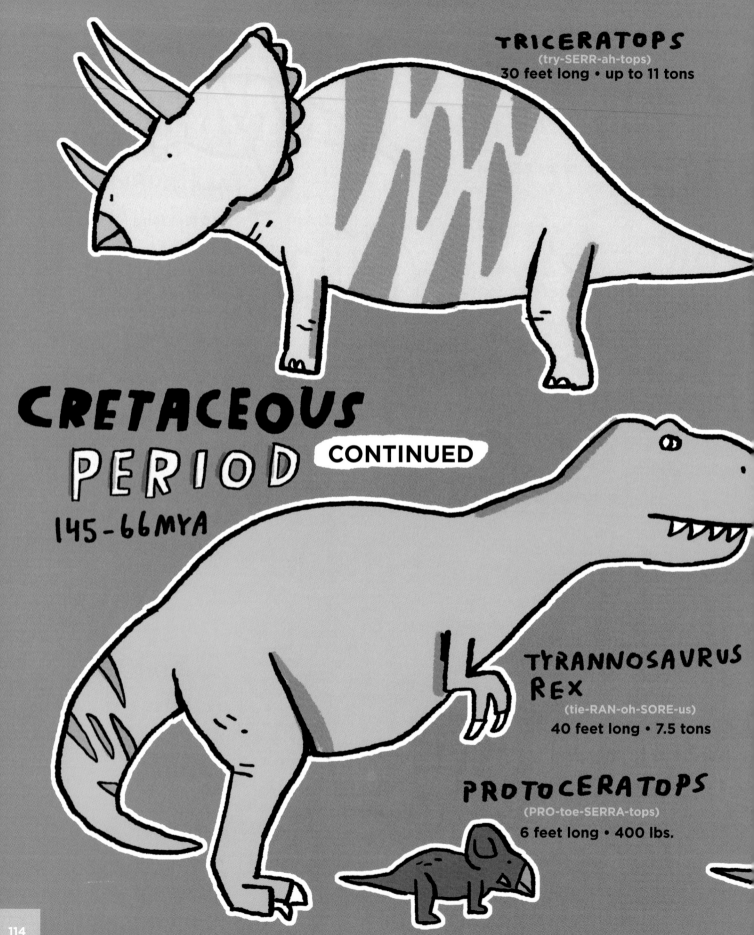

TRICERATOPS
(try-SERR-ah-tops)
30 feet long • up to 11 tons

CRETACEOUS
PERIOD CONTINUED
145-66MYA

TYRANNOSAURUS REX
(tie-RAN-oh-SORE-us)
40 feet long • 7.5 tons

PROTOCERATOPS
(PRO-toe-SERRA-tops)
6 feet long • 400 lbs.

OVIRAPTOR
(OH-vee-rap-tor)
8 feet long • 75 lbs.

VELOCIRAPTOR
(ve-LOSS-ih-rap-tor)
6 feet long • 33 lbs.

TROODON
(TROH-oh-don)
6.5 feet long • 110 lbs.

DEINONYCHUS
(die-NON-iih-cus)
13 feet long • 155 lbs.

ANKYLOSAURUS
(an-KILE-oh-SORE-us)
35 feet long • 7 tons

Kid for size comparison

PACHYCEPHALOSAURUS
(PACK-ee-KEFF-a-lo-SORE-us)
16 feet long • 2.5 tons

PRECAMBRIAN ERA

(4.6 BILLION YEARS AGO – 541 MILLION YEARS AGO)

THE EARLIEST PERIOD OF TIME ON EARTH.

ABOUT 90% OF EARTH'S HISTORY HAPPENED DURING THIS PERIOD.

OCEANS AND ROCKS FORMED.

VERY EARLY LIFE

SINGLE-CELLED ORGANISMS

DICKINSONIA

ONE OF THE FIRST ANIMALS

PALEOZOIC ERA

570 – 245 MILLION YEARS AGO

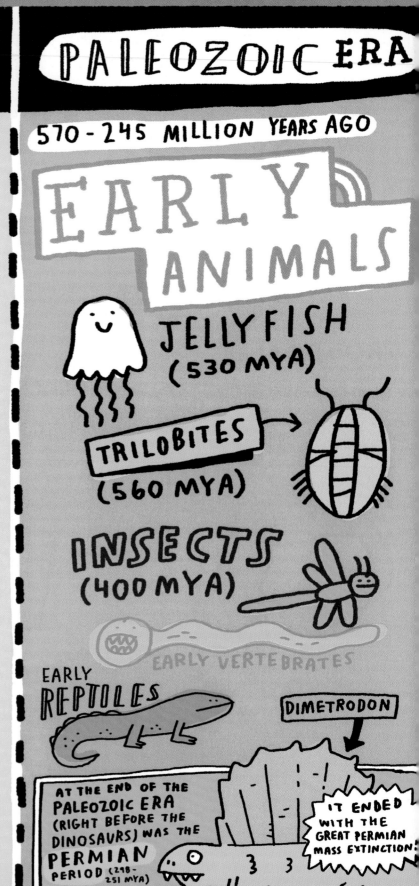

EARLY ANIMALS

JELLYFISH (530 MYA)

TRILOBITES (560 MYA)

INSECTS (400 MYA)

EARLY VERTEBRATES

EARLY REPTILES

DIMETRODON

AT THE END OF THE PALEOZOIC ERA (RIGHT BEFORE THE DINOSAURS) WAS THE **PERMIAN** PERIOD (298–251 MYA)

IT ENDED WITH THE GREAT PERMIAN MASS EXTINCTION!

TRIASSIC

PERIOD — 251-199 MYA — **YAY!**

CLIMATE: HOT AND DRY

THE **FIRST DINOSAURS** APPEARED.

HELLO

EARLY MAMMALS

DICYNODONT

EORAPTOR

VEGETATION LEPTOCYCAS

HORSETAILS

NOT ANOTHER MASS EXTINCTION!

AT THE END OF THE TRIASSIC, MANY PLANTS WENT EXTINCT AND 35% OF THE ANIMALS ON EARTH DIED!

MYA = MILLION YEARS AGO

← MESOZOIC ERA ─

JURASSIC

THE BIG DINO EXPLOSION!

PERIOD

199-145 MYA

FLYING REPTILES!

PLATE TECTONICS

GOOD BYE!

I'LL MISS YOU.

STEGOSAURUS

← BRACHIOSAURUS

CYCADS AND FERNS!

PLESIOSAURUS

CLIMATE: WARM AND HUMID

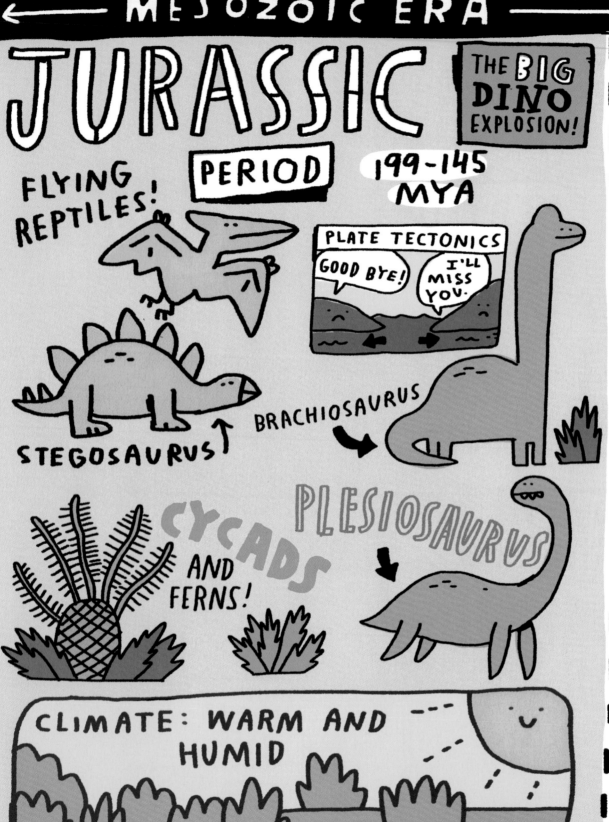

ALPHADON

BIRDS

BEES

FLOWERING PLANTS

HA HA!

HEE HEE!

WHAT DO YOU CALL A DINOSAUR THAT KNOWS **LOTS OF** WORDS?

A *THESAURUS!*

WHAT DOES A T. REX EAT?

ANYTHING IT WANTS!!!

WHY DID THE HERBIVORE CHEW UP THE FACTORY?

BECAUSE IT WAS A PLANT EATER.

WHICH DINOSAUR **NEVER GIVES** UP?

A *TRY*-CERETOPS!

DID YOU KNOW THAT I CAN JUMP HIGHER THAN A **HOUSE?!**

BECAUSE HOUSES CAN'T JUMP!

WHAT IS SOMETHING T. REXES HAD THAT NO OTHER ANIMAL **EVER** HAD?

BABY T. REXES!

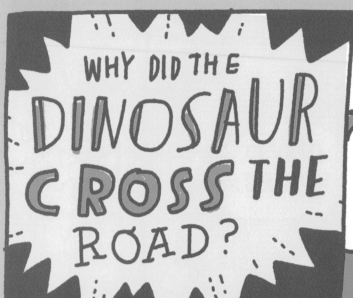

WHY DID THE **DINOSAUR CROSS THE** ROAD?

THE CHICKEN HADN'T EVOLVED YET!

I DON'T GET IT.

WHAT CAN A STEGOSAURUS **NEVER** HAVE FOR BREAKFAST?

YUMMY!

LUNCH AND DINNER!

BUT IT CAN HAVE PANGAEA CAKES!

WHAT DO YOU DO TO A BLUE TRICERATOPS?

YOU TRY TO CHEER IT UP!

SIGH

WHAT IS A DINO'S LEAST FAVORITE REINDEER?

COMET

Here are a few awesome books I devoured while working on this book!

Brusatte, Steve. *The Rise and Fall of the Dinosaurs: A New History of a Lost World*. New York, NY: William Morrow, 2018.

Cadbury, Deborah. *The Dinosaur Hunters: A True Story of Scientific Rivalry and the Discovery of the Prehistoric World*. New York, NY: HarperCollins, 2012.

Dixon, Dougal. *The Complete Illustrated Encyclopedia of Dinosaurs & Prehistoric Creatures*. London, UK: Southwater Books, 2014.

Lessem, Don. *Ultimate Dinopedia*. Washington, DC: National Geographic Children's Books, 2017.

Lloyd, Christopher. *The Nature Timeline Wallbook*. Kent, UK: What On Earth Books, 2017.

Pim, Keiron. *Dinosaurs — The Grand Tour*. New York, NY: The Experiment, 2016.

Richardson, Hazel. *Dinosaurs and Other Prehistoric Animals*. New York, NY: DK Publishing, 2003.

Woodward, John. *Dinosaur! Dinosaurs and Other Prehistoric Creatures As You've Never Seen Them Before*. New York, NY: DK Publishing, 2014.

I'd like to say a quick but very HUGE thank you to Katrin, the incredible team at Scholastic, and Josh Hathaway for all of their help with this book!
—M.L.

COMING SOON!